I0043626

Henry Neville Hutchinson

Extinct Monsters

A Popular Account of Some of the Larger Forms of Ancient Animal Life

Henry Neville Hutchinson

Extinct Monsters
A Popular Account of Some of the Larger Forms of Ancient Animal Life

ISBN/EAN: 9783744692069

Printed in Europe, USA, Canada, Australia, Japan

Cover: Foto ©berggeist007 / pixelio.de

More available books at **www.hansebooks.com**

EXTINCT MONSTERS.

A POPULAR ACCOUNT OF SOME OF THE LARGER
FORMS OF ANCIENT ANIMAL LIFE.

BY

REV. H. N. HUTCHINSON, B.A., F.G.S.,

AUTHOR OF "THE AUTOBIOGRAPHY OF THE EARTH,"
AND "THE STORY OF THE HILLS."

WITH ILLUSTRATIONS BY J. SMIT AND OTHERS.

THIRD THOUSAND, CORRECTED AND ENLARGED.

LONDON : CHAPMAN & HALL, LD.

1893.

All rights reserved.

"The possibilities of existence run so deeply into the extravagant that there is scarcely any conception too extraordinary for Nature to realise."—AGASSIZ.

PREFACE BY DR. HENRY WOODWARD, F.R.S.

KEEPER OF GEOLOGY, NATURAL HISTORY MUSEUM.

I HAVE been requested by my friend Mr. Hutchinson, to express my opinion upon the series of drawings which have been prepared by that excellent artist of animals, Mr. Smit, for this little book entitled " Extinct Monsters."

Many of the stories told in early days, of Giants and Dragons, may have originated in the discovery of the limb-bones of the Mammoth, the Rhinoceros, or other large animals, in caves, associated with heaps of broken fragments, in which latter the ignorant peasant saw in fancy the remains of the victims devoured at the monster's repasts.

In Louis Figuier's *World before the Deluge* we are favoured with several highly sensational views of extinct monsters; whilst the pen of Dr. Kinns has furnished valuable information as to the "slimy" nature of their blood !

The late Mr. G. Waterhouse Hawkins (formerly a lithographic artist) was for years occupied in unauthorised restorations of various Secondary reptiles and Tertiary mammals, and about 1853 he received encouragement from Professor Owen to undertake the restorations of extinct animals which still adorn the lower grounds of the Crystal Palace at Sydenham.

But the discoveries of later years have shown that the Dicynodon and Labyrinthodon, instead of being toad-like in form, were lacertilian or salamander-like reptiles, with elongated bodies and moderately long tails ; that the Iguanodon did not usually stand upon "all-fours," but more frequently sat up like some huge kangaroo with short fore limbs ; that the horn on its snout was really on its wrist ; that the Megalosaurus, with a more slender form of skeleton, had a somewhat similar erect attitude, and the habit, perhaps, of springing upon its prey, holding it with its powerful clawed hands, and tearing it with its formidable carnivorous teeth.

Although the Bernissart Iguanodon has been to us a complete revelation of what a Dinosaur really looked like, it is to America, and chiefly to the discoveries of Marsh, that we owe the knowledge of a whole series of new reptiles and mammals, many of which will be found illustrated within these pages.

Of long and short-tailed Pterodactyles we now know almost complete skeletons and details of their patagia or flying membranes. The discovery of the long-tailed feathered bird with teeth—the Archæopteryx, from the Oolite of Solenhofen, is another marvellous addition to our knowledge ; whilst Marsh's great Hesperornis, a wing-less diving bird with teeth, and his flying toothed bird, the Ichthyornis dispar, are to us equally surprising.

Certainly, both in singular forms of fossil reptilia and in early mammals, North America carries off the palm.

Of these the most remarkable are Marsh's Stegosaurus,

a huge torpid reptile, with very small head and teeth, about twenty feet in length, and having a series of flattened dorsal spines, nearly a yard in height, fixed upon the median line of its back ; and his Triceratops, another reptile bigger than Stegosaurus, having a huge neck-shield joined to its skull, and horns on its head and snout. Nor do the Eocene mammals fall short of the marvellous, for in Dinoceras we find a beast with six horns, and sword-bayonet tusks, joined to a skeleton like an elephant.

Latest amongst the marvels in modern palæontological discovery has been that made by Professor Fraas of the outline of the skin and fins in Ichthyosaurus tenuirostris, which shows it to have been a veritable shark-like reptile, with a high dorsal fin and broad fish-tail, so that "fish-lizard" is more than ever an appropriate term for these old Liassic marine reptiles.

As every palæontologist is well aware, restorations are ever liable to emendation, and that the present and latest book of extinct monsters will certainly prove no exception to the rule is beyond a doubt, but the author deserves our praise for the very boldness of his attempt, and the honesty with which he has tried to follow nature and avoid exaggeration. Every one will admire the simple and un-affected style in which the author has endeavoured to tell his story, avoiding, as far as possible, all scientific terms, so as to bring it within the intelligence of the unlearned. He has, moreover, taken infinite pains to study up his subject with care, and to consult all the literature bearing upon it. He has thus been enabled to convey accurate information in a simple and pleasing form, and to guide the artist in his difficult task with much wisdom and intelligence. That the excellence of the sketches is

b

due to the artist, Mr. Smit, is a matter of course, and so is the blame, where criticism is legitimate ; and no one is more sensible of the difficulties of the task than Mr. Smit himself.

Speaking for myself, I am *very well pleased* with the series of sketches ; and I may say so with the greater ease and freedom from responsibility, as I have had very little to do with them, save in one or two trifling matters of criticism. I may venture, however, to commend them to my friends among the public at large as the happiest set of restorations that has yet appeared.

H. W.

THE LATE SIR RICHARD OWEN AND A SKELETON OF DINORNIS MAXIMUS.

PLATE XXIV. (From a photograph.)

AUTHOR'S PREFACE.

NATURAL history is deservedly a popular subject. The manifestations of life in all its varied forms is a theme that has never failed to attract all who are not destitute of intelligence. From the days of the primitive cave-dwellers of Europe, who lived with mammoths and other animals now lost to the world; of the ancient Egyptians, who drew and painted on the walls of their magnificent tombs the creatures inhabiting the delta of the Nile; of the Greeks, looking out on the world with their bright and child-like curiosity, down to our own times, this old, yet ever new, theme has never failed. Never before was there such a profusion of books describing the various forms of life inhabiting the different countries of the globe, or the rivers, lakes, and seas that diversify its scenery. Popular writers have done good service in making the way plain for those who wish to acquaint themselves with the structures, habits, and histories of living animals; while for students a still greater supply of excellent manuals and text-books has been, and still continues to be, forthcoming.

But in our admiration for the present we forget the great past. How seldom do we think of that innumerable host of creatures that once trod this earth! How little in comparison has been done for *them !* Our natural-history books deal only with those that are alive now. Few popular writers have attempted to depict, as on a canvas, the great earth-drama that has, from age to age, been enacted on the terrestrial stage, of which we behold the latest, but probably not the closing scenes.

When our poet wrote "All the world's a stage," he thought only of "men and women," whom he called "merely players," but the geologist sees a wider application of these words, as he reviews the drama of past life on the globe, and finds that animals, too, have had "their exits and their entrances;" nay more, "the strange eventful history" of a human life, sketched by the master-hand, might well be chosen to illustrate the birth and growth of the tree of life, the development of which we shall briefly trace from time to time, as we proceed on our survey of the larger and more wonderful animals of life that flourished in bygone times.

We might even make out a "seven ages" of the world, in each of which some peculiar form of life stood out prominently, but such a scheme would be artificial.

There is a wealth of material for reconstructing the past that is simply bewildering; and yet little has been done to bring before the public the strange creatures that have perished.[1]

To the writer it is a matter of astonishment that the

---

[1] Figuier's *World before the Deluge* is hardly a trustworthy book, and is often not up to date. The restorations also are misleading. Professor Dawson's *Story of the Earth and Man* is better; but the illustrations are poor. Nicholson's *Life-History of the Earth* is a student's book. Messrs. Cassells' *Our Earth and its Story* deals with the whole of geology, and so is too diffusive; its ideal landscapes and restorations leave much to be desired.

discoveries of Marsh, Cope, Leidy, and others in America, not to mention some important European discoveries, should have attracted so little notice in this country. In the far and wild West a host of strange reptiles and quadrupeds have been unearthed from their rocky sepulchres, often of incredibly huge proportions, and, in many cases, more weird and strange than the imagination could conceive; and yet the public have never heard of these discoveries, by the side of which the now well-known "lost creations" of Cuvier, Buckland, or Conybeare sink into the shade. For once, we beg leave to suggest, the hungry pressman, seeking "copy," has failed to see a good thing. Descriptions of some of "Marsh's monsters" and how they were found, might, one would think, have proved attractive to a public ever on the look out for something new.

Professor Huxley, comparing our present knowledge of the mammals of the Tertiary era with that of 1859, states that the discoveries of Gaudry, Marsh, and Filhol, are "as if zoologists were to become acquainted with a country hitherto unknown, as rich in novel forms of life as Brazil or South America once were to Europeans."

The object of this book is to describe some of the larger and more monstrous forms of the past—the lost creations of the old world; to clothe their dry bones with flesh, and suggest for them backgrounds such as are indicated by the discoveries of geology: in other words, to endeavour, by means of pen and pencil, to bring them back to life. The ordinary public cannot learn much by merely gazing at skeletons set up in museums. One longs to cover their nakedness with flesh and skin, and to see them as they were when they walked this earth.

Our present imperfect knowledge renders it difficult in some cases to construct successful restorations ; but, nevertheless, the attempt is worth making : and if some who think geology a very dry subject, can be converted to a different opinion on reading these pages, we shall be well rewarded for our trouble.

We venture to hope that those who will take the trouble to peruse this book, or even to look at its pictures, on which much labour and thought have been expended, will find pleasure in visiting the splendid geological collection at Cromwell Road. We have often watched visitors walking somewhat aimlessly among those relics of a former world, and wished that we could be of some service. But, if this little book should help them the better to understand what they see there, our wish will be accomplished.

Another object which the writer has kept in view is to connect the past with the present. It cannot be too strongly urged that the best commentary on the dead past is the living present. It is unfortunate that there is still too great a tendency to separate, as by a great gulf, the dead from the living, the past from the present, forms of life. The result of this is seen in our museums. Fossils have too often been left to the attention of geologists not always well acquainted with the structures of living animals. The more frequent introduction of fossil specimens side by side with modern forms of life would not only be a gain to the progress and spread of geological science, but would be a great help to students of anatomy and natural history. The tree of life is but a mutilated thing, and half its interest is gone, when the dead branches are lopped off.

It is, perhaps, justifiable to give to the term " monster " a somewhat extended meaning. The writer has therefore included in his menagerie of extinct animals one or two creatures which, though not of any great size, are nevertheless remarkable in various ways—such, for instance, as the winged reptiles, and anomalous birds with teeth, of later times, and others. Compared with living forms, these creatures appear to us as "monstrosities," and may well find a place in our collection.

The author wishes, in a few words, to thank those friends who have rendered him assistance in his task.

Dr. Henry Woodward, F.R.S., Keeper of Geology, Natural History Museum, has from the first taken a lively interest in this little book. He kindly helped the author with his advice on difficult matters, criticising some of the artist's preliminary sketches and suggesting improvements in the restorations. With unfailing courtesy he has ever been willing, in spite of many demands on his time, to place his knowledge at the disposal of both the author and artist ; and in this way certain errors have been avoided. Besides this, he took the trouble to read through the proof-sheets, and made suggestions and corrections which have greatly improved the text. For all this welcome aid the author begs to return his sincere thanks.

To Mr. Smith Woodward, of the Natural History Museum, the author is also much indebted for his kindness in reading through the text and giving valuable information with regard to the latest discoveries.

The artist, Mr. Smit, notwithstanding the novelty of the subject and the difficulties of the task, has thrown himself heartily into the work of making the twenty-four restorations of extinct animals. To him, also, the author is greatly indebted, and considers himself fortunate in having secured the services of so excellent an artist.

To the publishers his thanks are due for their liberality in the matter of illustrations, and the readiness with which they have responded to suggestions.

With regard to minor illustrations the following acknowledgments are due :—

To the Palæontological Society of Great Britain for permission to reproduce three of the illustrations in Sir Richard Owen's great work, *British Fossil Reptiles*, published in their yearly volumes, viz. Figs. 3, 4, and 8.

To Messrs. Bell and Co. for the following cuts from the late Dr. Gideon A. Mantell's works : viz. Figs. 12, 14, 20, 33, 37, 38.

To Messrs. A. and C. Black for the following cuts from Owen's *Palæontology :* viz. Figs. 51, 54, 56, 57.

Appendix IV. contains a list of some of the works of which the writer has made use ; but it would be impossible within reasonable limits to enumerate all the separate papers which have necessarily been consulted. The reader will find numerous references, such as " Case Y on Plan," in brackets; these refer to the plan given at the end of the excellent little *Guide to the Exhibition Galleries in the Department of Geology and Palæontology in the Natural History Museum*, Cromwell Road (price one shilling), which visitors to the Museum are advised to obtain.

PREFACE TO SECOND EDITION.

THE appearance of a second edition affords the author a pleasant opportunity of thanking the reading public, and the Press, for the kind way in which his endeavour to popularise the results of modern Palæontology has been received. There seem to be fashions in all things—even in sciences ; and perhaps the wonderful advances we have witnessed of late years in the physical sciences on the one hand, and in biological sciences on the other, may have tended to throw Palæontology somewhat into the shade. Let us hope that it will not remain there long.

A large number of illustrations have been added for the present edition, besides additional matter here and there in the text. Three of the plates (viz. Plates II. X. XV.) have been redrawn. Plate II. shows the Ichthyosaurus as interpreted by the latest discovery from Würtemburg. Plate X. gives a somewhat different interpretation of the Stegosaurus, suggested by some remarks of Mr. Lydekker.

A slight change will be noticed in Plate XV. (Brontops). Plate XVII. is a great improvement on the old drawing (Fig. 28, old edition) of the Megatherium skeleton. Plate XXIV., besides containing a valuable portrait of the late Sir Richard Owen, gives another drawing of the Dinornis skeleton.

April, 1893.

CONTENTS.

LIST OF FULL-PAGE ILLUSTRATIONS.

LIST OF FIGURES IN TEXT.

EXTINCT MONSTERS.

INTRODUCTION.

" The earth hath gathered to her breast again
And yet again, the millions that were born
Of her unnumbered, unremembered tribes."

LET us see if we can get some glimpses of the primæval inhabitants of the world, that lived and died while as yet there were no
men and women having authority over the fishes of the sea and
the fowls of the air.

We shall, perhaps, find this antique world quite as strange as
the fairy-land of Grimm or Lewis Carroll. True, it was not
inhabited by "slithy toves" or "jabber-wocks," but by real
beasts, of whose shapes, sizes, and habits much is already known
—a good deal more than might at first be supposed. And yet,
real as it all is, this antique world—this panorama of scenes that
have for ever passed away—is a veritable fairy-land. In those
days of which geologists tell us, the principal parts were played,
not by kings and queens, but by creatures many of which
were very unlike those we see around us now. And yet it is no
fairy-land after all, where impossible things happen, and where
impossible dragons figure largely; but only the same old
world in which you and I were born. Everything you will see
here is quite true. All these monsters once lived. Truth is

B

stranger than fiction; and perhaps we shall enjoy our visit to this fairy-land all the more for that reason. For not even the dragons supposed to have been slain by armed knights in old times, when people gave ear to any tale, however extravagant, could equal in size or strength the real dragons we shall presently meet with, whose actual bones may be seen in the Natural History Museum at South Kensington.

Many people who visit this great museum and find their way to the geological galleries on the right, pass hastily by the cases of bones, teeth, and skeletons. These things, it seems, fail to interest them. They do not know how to interpret them. They cannot picture to themselves the kinds of creatures to which the relics once belonged; and so they pass them by and presently go to the more attractive collection of stuffed birds on the other side. There they see the feathered tribes of the air all beautifully arranged; some poised in the air by almost invisible wires; some perched on branches : but all surrounded by grass, flowers, and natural objects, imitated with marvellous reality, so that they see the birds as they really are in nature, and can almost fancy they hear them singing.

Now, it has often occurred to the present writer that something more might be done for the long-neglected "lost creations" of the world, to bring them out of their obscurity, that they may be made to tell to the passer-by their wondrous story. We can, however, well imagine some of our readers asking, "Can these dry bones live?" "Yes," we would say, "they can be made to live; reason and imagination will, if we give them proper play, provide us eyes wherewith to see the world's lost creations." To such men as Cuvier, Owen, Huxley, and others, these dry bones *do* live. It will be our object to describe to the reader some of the wonderful results that have rewarded the lifelong labours of such great men. We shall take some of the largest and strangest forms of life that once lived, and try to picture them as they really were when alive, whether walking on land, swimming in the sea, or flying in the air; to understand the meanings of their more obvious structures ; and to form some conclusions with regard to their habits, as well as to find out, if possible, their relations,—as far as such questions have been answered by those most qualified to settle these difficult matters.

All technical details, such as the general reader is unfamiliar with, will be as far as possible suppressed. Let us fancy a long procession of extinct monsters passing in single file before us, and ourselves endeavouring to pick out their "points" as they present themselves to the eye of imagination. It is not, be it remembered, mere imagination that guides the man of science in such matters, for all his conclusions are carefully based on reason ; and when conclusions are given, we shall endeavour to show how they have been arrived at.

For millions of years countless multitudes of living animals have played their little parts on the earth and passed away, to be buried up in the oozy beds of the seas of old time, or entombed with the leaves that sank in the waters of primæval lakes. The majority of these perished beyond all recovery, leaving not a trace behind ; yet a vast number of fossilised remains have been, in various ways, preserved; sometimes almost as completely as if Dame Nature had thoughtfully embalmed them for our instruction and delight.

Down in those old seas and lakes she kept her great museum, in order to preserve for us a selection of her treasures. In course of time she slowly raised up sea-beds and lake-bottoms to make them into dry land. This museum is everywhere around us. We have but to enter quarries and railway cuttings, or to search in coal-mines, or under cliffs at the seaside, and we can consult her records. As the ancient Egyptians built tombs, pyramids, and temples, from which we may learn their manner of life and partly read their history, so Nature has entombed, not one race only, but many races of the children of life. Her records are written in strange hieroglyphs, yet it is not difficult to interpret their meaning; and thus many an old story, many an old scene, may be pictured in the mind of man.

Shall we call this earth-drama a tragedy or a comedy? Doubtless tragic scenes occurred at times; as, for instance, when fierce creatures engaged in deadly combat: and probably amusing, if not comic, incidents took place occasionally, such as might have provoked us to laughter, had we been there to see them. But let us simply call it a drama. Backgrounds of scenery were not wanting. Then, as now, the surface of the earth was clothed with vegetation, and strange cattle pastured on grassy plains. Vegetation was at times very luxuriant. The forests of the coal period, with their giant reeds and club-moss trees, must have made a strange picture. Then, as now, there rose up from the plains lofty ranges of mountains, reaching to the clouds, their summits clothed with the eternal snows. These, too, played their part, feeding the streams and the rivers that meandered over the plains, bringing life and fertility with them, as they do now. The sun shone and the wind blew : sometimes gently, so that the leaves just whispered in an evening breeze; at other times so violently that the giants of the forest swayed to and fro, and the seas lashed themselves furiously against rocky coasts. Nor were the underground forces of the earth less active than they are now : volcanic eruptions often took place on a magnificent scale ; volcanoes poured out fiery lava streams for leagues beneath their feet; great showers of ashes and fine dust were ejected in the air, so that the sun was darkened for a time, and the surface of the sea was covered for many miles with floating pumice and volcanic dust, which in time sank to the bottom, and was made into hard rock, such as we now find on the top of Snowdon.

Earthquake shocks were quite as frequent, and no doubt the ground swayed to and fro, or was rent open as some unusually great earth-movement took place, and perhaps a mountain range was raised several feet or yards higher. All this we learn from the testimony of the rocks beneath our feet. It only requires the use of a little imagination to conjure up scenes of the past, and paint them as on a moving diorama.

We shall not, however, dwell at any length on the scenery, or the vegetation that clothed the landscape at different periods; for these features are sufficiently indicated in the beautiful drawings of extinct animals by our artist, Mr. J. Smit.

The researches of the illustrious Baron Cuvier, at Paris, as embodied in his great work, *Ossemens Fossiles*, gave a great impetus to the study of organic remains. It was he who laid the foundations of the science of Palæontology,[1] which, though much has already been accomplished, yet has a great future before it. Agassiz, Owen, Huxley, Marsh, Cope, and others, following in his footsteps, have greatly extended its boundaries; but he was the pioneer.

Before his time fossil forms were very little known, and still less understood. His researches, especially among vertebrates, or back-boned animals, revealed an altogether undreamed-of wealth of entombed remains. It is true the old and absurd notion that fossils were mere "sports of Nature," sometimes bearing more or less resemblance to living animals, but still only an accidental (!) resemblance, had been abandoned by Leibnitz, Buffon, and Pallas; and that Daubenton had actually compared the fossil bones of quadrupeds with those of living forms; while Camper declared his opinion that some of these remains belonged to extinct species of quadrupeds.

It is to Cuvier, however, that the world owes the first systematic application of the science of comparative anatomy, which he himself had done so much to place on a sound basis, to the study of the bones of fossil animals. He paid great attention to

---

[1] Palæontology is the science which treats of the living beings, whether animal or vegetable, which have inhabited this globe at past periods in its history. (Greek—*palaios*, ancient; *onta*, beings; *logos*, discourse.)

the relative shapes of animals, and the different developments of the same kind of bones in various animals, and especially to the nature of their teeth. So great did his experience and knowledge become, that he rarely failed in naming an animal from a part of its skeleton. He appreciated more clearly than others before him the mutual dependence of the various parts of an animal's organisation. "The organism," he said, "forms a connected unity, in which the single parts cannot change without modifications in the other parts."

It will hardly be necessary to give examples of this now well-known truth; but, just to take one case: the elephant has a long proboscis with which it can reach the ground, and consequently its neck is quite short; but take away the long proboscis, and you would seriously interfere with the relation of various parts of its structure to each other. How, then, could it reach or pick up anything lying on the ground? Other changes would have to follow: either its legs would require to be shortened, or its neck to be lengthened. In every animal, as in a complex machine, there is a mutual dependence of the different parts.

As he progressed in these studies, Cuvier was able with considerable success to restore extinct animals from their fossilised remains, to discover their habits and manner of life, and to point out their nearest living ally. To him we owe the first complete demonstration of the possibility of restoring an extinct animal. His "Law of Correlation" however, has been found to be not infallible; as Professor Huxley has shown, it has exceptions. It expresses our experience among living animals, but, when applied to the more ancient types of life, is liable to be misleading.

To take one out of many examples of this law : Carnivorous animals, such as cats, lions, and tigers, have claws in their feet, very different from the hoofs of an ox, which is herbivorous ; while the teeth of the former group are very different to those of the latter. Thus the teeth and limbs have a certain definite relation to each other, or, in other words, are correlated. Again, horned quadrupeds are all herbivorous (or graminivorous), and have hoofs to their feet. The following amusing anecdote serves to illustrate Cuvier's law. One of his students thought he would try and frighten his master, and, having dressed up as a wild beast, entered Cuvier's bedroom by night, and, presenting himself by his bedside, said in hollow tones, "Cuvier, Cuvier, I've come to eat you!" The great naturalist, who on waking up was able to discern something with horns and hoofs, simply remarked, "What! horns, hoofs—graminivorous—you can't!" What better lesson could the master have given the pupil to help him to remember his "Law of Correlation"?

Cuvier's great work, entitled *Ossemens Fossiles*, will long remain an imperishable monument of the genius and industry of the greatest pioneer in this region of investigation. This work proved beyond a doubt to his astonished contemporaries the great antiquity of the tribes of animals now living on the surface of the earth. It proved more than that, however; for it showed the existence of a great philosophy in Nature which linked the past with the present in a scheme that pointed to a continuity of life during untold previous ages. All this was directly at variance with the prevalent ideas of his time, and consequently his views were regarded by many with alarm, and he received a good deal of abuse—a fate which many other original thinkers before him have shared.

It is somewhat difficult for people living now, and accustomed to modern teaching, to realise how novel were the conclusions announced by Cuvier. In his *Discourse on the Revolutions of the Surface of the Globe*, translated into most European languages under the title *Theory of the Earth*, he lays down, among others, the two following propositions :—

1. That all organised existences were not created at the same time; but at different times, probably very remote from each

other—vegetables before animals, mollusca and fishes before reptiles, and the latter before mammals.

2. That fossil remains in the more recent strata are those which approach nearest to the present type of corresponding living species.

Teaching such as this gave a new impetus to the study of organic remains, and Palæontology, as a science, began with Cuvier.

CHAPTER I.

HOW EXTINCT MONSTERS ARE PRESERVED.

"Geology, beyond almost every other science, offers fields of research adapted to all capacities and to every condition and circumstance of life in which we may be placed. For while some of its phenomena require the highest intellectual powers, and the greatest attainments in abstract science for their successful investigation, many of its problems may be solved by the most ordinary intellect, and facts replete with the deepest interest may be gleaned by the most casual observer."—MANTELL.

LET us suppose we are visiting a geological museum for the first time, passing along from one department to another with ever-increasing wonder—now admiring the beautiful polished marbles from Devonshire, with their delicate corals, or the wonderful fishes from the Old Red Sandstone, with their plates of enamel; now the delicate shells and ammonites from the Lias or Oolites, with their pearly lustre still preserved; now the white fresh-looking shells from the Isle of Wight; now the ponderous bones and big teeth of ancient monsters from the Wealden beds of Sussex. The question might naturally occur, "How were all these creatures preserved from destruction and decay, and sealed up so securely that it is difficult to believe they are as old as the geologists tell us they are?" It will be worth our while to consider this before we pass on to describe the creatures themselves.

Now, in the first place, "fossils" are not always "petrifactions," as some people seem to think; that is to say, they are not all turned into stone. This is true in many cases, no doubt, yet one frequently comes across the remains of plants and animals that have undergone very little change, and have, as it were, been simply sealed up. The state of a fossil depends on several circumstances, such as the soil, mud, or other medium in which it may happen to be preserved. Again, the newest, or most recent, fossils are generally the least altered. We have fossils of all ages, and in all states of preservation. As examples of fossils very little altered, we may take the case of the wonderful collection of bones discovered by Professor Boyd Dawkins in caves in various parts of Great Britain. The results of many years of research are given in his most interesting book on *Cave-Hunting*. This enthusiastic explorer and geologist has discovered the remains of a great many animals, some of which are quite extinct, while others are still living in this country. These remains belong to a late period, when lions, tigers, cave-bears, wolves, hyænas, and reindeer inhabited our country. In some cases the caves were the dens of hyænas, who brought their prey into caverns in our limestone rocks, to devour them at their leisure; for the marks of their teeth may yet be seen on the bones. In other cases the bones seem to have been washed into the caves by old streams that have ceased to run; but in all cases they are fairly fresh, though often stained by iron-rust brought in by water that has dissolved iron out of various rocks—for iron is a substance met with almost everywhere in nature. Sometimes they are buried up in a layer of soil, or "cave-earth," and at other times in a layer of stalagmite—a deposit of carbonate of lime gradually formed on the floors of caves by the evaporation of water charged with carbonate of lime.

Air and water are great destroyers of animal and vegetable substances from which life has departed. The autumn leaves that fall by the wayside soon undergo change, and become at last separated or resolved into their original elements. In the same way when any wild animal, such as a bird or rabbit, dies in an exposed place, its flesh decays under the influence of rain and wind, so that before long nothing but dry bones is left. Hamlet's wish that this "too too solid flesh would melt" is soon realised after death ; and that active chemical element in the air known as oxygen, in breathing which we live, has a tenfold power over dead matter, slowly causing chemical actions somewhat similar to those that take place in a burning candle, whereby decaying flesh is converted into water-vapour and carbonic acid gas. Thus we see that oxygen not only supports life, but breaks up into simpler forms the unwholesome and dangerous products of decaying matter, thus keeping the atmosphere sweet and pure ; but in time, even the dry bones of the bird or rabbit, though able for a longer period to resist the attacks of the atmosphere, crumble into dust, and serve to fertilise the soil that once supported them.

Now, if water and air be excluded, it is wonderful how long even the most perishable things may be preserved from this otherwise universal decay. In the Edinburgh museum of antiquities may be seen an old wooden cask of butter that has lain for centuries in peat—which substance has a curiously preservative power ; and human bodies have been dug out of Irish peat with the flesh well preserved, which, from the nature of the costume worn by the person, we can tell to be very ancient. Meat packed in tins, so as to be entirely excluded from the air, may be kept a very long time, and will be found to be quite fresh and fit for use.

But air and water have a way of penetrating into all sorts of places, so that in nature they are almost everywhere. Water can slowly filter through even the hardest rocks, and since it contains dissolved air, it causes the decay of animal or vegetable substances. Take the case of a dead leaf falling into a lake, or some quiet pool in a river. It sinks to the bottom, and is buried up in gravel, mud, or sand. Now, our leaf will stand a very poor chance of preservation on a sandy or gravelly bottom, because these materials, being porous, allow the water to pass through them easily. But if it settles down on fine mud it may be covered up and become a fossil. In time the soft mud will harden into clay or shale, retaining a delicate impression of the leaf; and even after thousands of years, the brown body of the leaf will be there, only partly changed. In the case of the plants found in coal, the lapse of ages since they were buried up has been so great (and the strata have been so affected by the great pressure and by the earth's internal heat) that certain chemical changes have converted leaves and stems into carbon and some of its compounds, much in the same way that, if you heat wood in a closed vessel, you convert it into charcoal, which is mostly carbon. The coal we burn in our fires is entirely of vegetable origin, and every seam in a coal-mine is a buried forest of trees, ferns, reeds, and other plants.

The reader will understand how it is that rocks composed of hardened sand or gravel, sandstones and conglomerates, contain but few fossils; while, on the other hand, such rocks as clay, shale, slate, and limestone often abound in fossils, because they are formed of what was once soft mud, that sealed up and protected corals, shell-fish, sea-urchins, fishes, and other marine animals. Had they been covered up in sand the chances are that percolating water would have slowly dissolved the shells and corals, the hard coats of the crabs, and the bones of the fishes, all of which are composed of carbonate of lime ; and we know that is a substance easily dissolved by water.

It is in the rocks formed during the later geological periods that we find fossils least changed from their original state ; for time works great changes, and too little time has elapsed since those periods for any considerable alterations to have taken place. But when we come to examine some of the earlier rocks, which have been acted upon in various ways for long periods of time, such as the pressure of vast piles of overlying rocks, and the percolation of water charged with mineral substances (water sometimes warmed by the earth's internal heat), then we may expect to find the remains of the world's lost creations in a much more mineralised condition. Every fossil-collector must be familiar with examples of changes of this kind. For instance, shells originally composed of carbonate of lime are often found to have been turned into flint or silica. Another curious change is illustrated in the case of a stratum found in Cambridgeshire and other counties. In this remarkable layer, only about a foot in thickness, one frequently finds bones and teeth of fishes and reptiles. These, however, have all undergone a curious change, whereby they have been converted into phosphate of lime—a compound of phosphorus and lime. It abounds in "nodules," or lumps, of this substance, which, along with thousands of fossils, are every year ground up and converted by a chemical process into valuable artificial manure for the farmer.

The soft parts of animals, as we have said before, cannot be preserved in a fossil state; but, as if to compensate for this loss, we sometimes meet with the most faithful and delicate impressions. Thus, cuttle-fishes have, in some instances, left, on the clays which buried them up, impressions of their soft, long arms, or tentacles, and, as the mud hardened into solid rock, the impressions are fixed imperishably. Examples of these interesting records may be seen at the Natural History Museum at South Kensington. Even soft jelly-fishes have left their mark on certain rocks! At a place in Bavaria, called Solenhofen, there is a remarkably fine-grained limestone containing a multitude of wonderful impressions. This stone is well known to lithographers, and is largely used in printing. On it the oldest known bird has left its skeleton and faithful impressions of its feathers.

The footprints of birds and reptiles are by no means uncommon. Such records are most valuable, for a great deal may be learned from even a footprint as to the nature of the animal that made it (see p. 9).

Since the greater number of animals described in this book are reptiles, quadrupeds, and other inhabitants of the land, and only a few had their home in the sea, we must endeavour to try and understand how their remains may have been preserved. Our object in writing this book is to interpret their story, and, as it were, to bring them to life again. Each one must be made to tell its own story, and that story will be far from complete if we cannot form some idea of how it found its way into a watery grave, and so was added to Nature's museum. For this purpose we must briefly explain to the reader how the rocks we see around us have been deposited; for these rocks are the tombs in which lost creations lie.

Go into any ordinary quarry, where the men are at work, getting out the stone in blocks to be used in building, or for use on the roads, or for some other purpose, and you will be pretty sure to notice at the first glance that the rock is arranged as if it had been built up in layers. Now, this is true of all rocks that have been laid down by the agency of water—as most of them have been. True, there are exceptions, but every rule has its exceptions. If you went into a granite quarry at Aberdeen, or a basalt quarry near Edinburgh, you would not see these layers; but such rocks as these do not contain fossils. They have been mainly formed by the action of great heat, and were forced up to the surface of the earth by pressure from below. As they slowly cooled, the mineral substances of which they were formed gradually crystallised; and it is this crystalline state, together with the signs of movement, that tells us of their once heated state. Such rocks are said to be of igneous origin (Lat. *ignis*, fire). But nearly all the other rocks were formed by the action of water—that is, under water,—and hence are known to geologists as aqueous deposits (Lat. *aqua*, water). They may be considered as sediments that slowly settled down in seas, lakes, or at the mouths of rivers. Such deposits are in the course of being formed at the present day. All round our coasts mud, sand, and gravel are being accumulated, layer by layer. These materials are constantly being swept off the land by the action of rain and rivers, and carried down to the sea. Perhaps, when staying at the sea-side, you may have noticed, after rainy and rough weather, how the sea, for some distance from the shore, is discoloured with mud—especially at the mouth of a river. The sand, being heavy, soon sinks down, and this is the reason why sand-bars so frequently block the entrance to rivers. Then again, the waves of the sea beat against the sea-shore and undermine the cliffs, bringing down great fragments, which. after a time are completely broken up and worn down into rounded pebbles, or even fine sand and mud. It is very easy to see that in this way large quantities of sand, gravel, and mud are continually supplied to our seas. We can picture how they will settle down ; the sand not far from the shore, and the fine mud further out to sea. When the rough weather ceases, the river becomes smaller and flows less rapidly, so that when the coarse *débris* of the land has settled down to form layers, or strata, of sand and gravel, then the fine mud will begin to settle down also, and will form a layer overlying them or further out. Thus we learn, from a little observation of what is now going on, how layers of sand and mud, such as we see in a quarry, were made thousands and thousands of years ago.

When we think of all the big rivers and small streams continually flowing into the sea, we shall begin to realise what a great work rain and rivers are doing in making the rocks of the future. If, at a later period, a slight upheaval of the sea-bed were to take place so as to bring it above water, and such is very likely, these materials would be found neatly arranged in layers, and more or less hardened into solid rock.

The reader may, perhaps, find it rather hard at first to realise that in this simple way vast deposits of rock are being formed in the seas of the present day, and that the finer material thus derived from a continent may be · carried by ocean currents to great distances ; but so it is. Over thousands of square miles of ocean, deposits are being gradually accumulated which will doubtless be some day turned into hard rock. Just to take one example : it has been found that in the Atlantic Ocean, a distance of over two hundred miles from the mouth of that great river, the Amazon, the sea is discoloured by fine sediment.

There is another kind of rock frequently met with, the building up of which cannot be explained in the way we have pointed out; and that is limestone. This rock has not been deposited as a sediment, like clays and sandstones, but geologists have good reasons for believing that it has been gradually formed in the deeper and clearer parts of oceans by the slow accumulation of marine shells, corals, and other creatures, whose bodies are partly composed of carbonate of lime. This seems incredible at first, but the proofs are quite convincing.[1] As Professor Huxley well remarked, there is as good evidence that chalk has been built up by the accumulation of minute shells as that the Pyramids were built by the ancient Egyptians.

The science of geology reveals the startling fact that all the great series of the stratified rocks, whose united thickness is over 80,000 feet, has been mainly accumulated under water, either by the action of those powerful geological agents—rain and rivers—or through the agency of myriads of tiny marine animals. When we have grasped this idea, we have learned our first, and, perhaps, most useful lesson in geology.

Now let us apply what has been above explained to the question immediately before us. We want to know how the skeletons of animals living on land came to be buried up under water, among the stratified rocks that are to be seen all over our country, and most of which were made under the sea.

We can answer this question by going to Nature herself, in order to find out what is actually going on at the present time, by inquiring into the habits of land animals, their surroundings, and the accidents to which they are liable at sundry times and in divers manners. It is by this simple method of studying present actions that nearly all difficult questions in geology may be solved. The leading principle of the geologist is to interpret the past by

[1] See *The Autobiography of the Earth*, p. 223.

the light of the present, or, in other words, to find out what happens now, in order to learn what took place ages ago ; for it is clear that the world has been going on in the same way for at least as far back as geological history can take us. There has been a *uniformity*, or sameness, in Nature's actions ever since living things first dwelt on the earth.

Just as rivers are mainly responsible for bringing down to the sea the materials of which rocks are made, so these universal carrying agents are the means by which the bodies of many animals that live in the plains, over which they wander, are brought to their last resting-place. We have only to consult the records of great floods to see what fearful havoc they sometimes make among living things, and how the dead bodies are swept away.

Great floods rise rapidly, so that the herds of wild animals pasturing on grassy plains are surprised by the rising waters, and, being unable to withstand the force of the water, are hurried along, and so drowned. When dead they sink to the bottom, and may, in some cases, be buried up in the *débris* hurried along by the river ; but as a rule their bodies, being swollen by the gases formed by decomposing flesh, rise again to the surface, and consequently may be carried along for many a mile, till they reach some lake, or perhaps right down to the mouth of a river, and so may be taken out to sea.

One or two examples will be given to show how important is the action of such floods. Sir Charles Lyell has given some striking illustrations of this. There was a memorable flood in the southern borders of Scotland on the 24th of June, 1794, which caused great destruction in the region of the Solway Firth. Heavy rains had fallen, so that every stream entering the firth was greatly swollen. Not only sheep and cattle, but even herdsmen and shepherds were drowned. When the flood had subsided a fearful spectacle was seen on a large sand-bank, called "the beds of Esk," where the waters meet ; for on this one bank were found collected together the bodies of 9 black cattle, 3 horses, 1840 sheep, 45 dogs, 180 hares, together with those of many smaller animals, also the corpses of two men and one woman.

Humboldt, the celebrated traveller, says that when, at certain seasons, the large rivers of South America are swollen by heavy rains, great numbers of quadrupeds are drowned every year. Troops of wild horses that graze in the "savannahs," or grassy plains, are said to be swept away in thousands.

In Java, in the year 1699, the Batavian River was flooded during an earthquake, and drowned buffaloes, tigers, rhinoceroses, deer, apes, crocodiles, and other wild beasts, which were brought down to the coast by the current.

In tropical countries, where very heavy rains fall at times, and rivers become rapidly swollen, floods are a great source of danger to man and beast. Probably the greater number of the bodies of animals thus drowned find their way into lakes, through which rivers flow, and never reach the sea ; and if the growth of sediment in such lakes goes on fairly rapidly, their remains may be buried up, and so preserved. But in many cases the bones fall one by one from the floating carcase, and so may in that way be scattered at random over the bottom of the lake, or the bed of a river at its mouth. In hot countries such bodies, on reaching the sea, run a great chance of being instantly devoured by sharks, alligators, and other carnivorous animals. But during very heavy floods, the waters that reach the sea are so heavily laden with mud, that these predaceous animals are obliged to retire to some place where the waters are clear, so that at such times the dead bodies are more likely to escape their ravages; and, at the same time, the mud with which the waters are charged falls so rapidly that it may quickly cover them up. We shall find further on that this explanation probably applies to the case of the "fish-lizards," whose remains are found in the Lias formation (see p. 51).

But, for several reasons, sedimentary rocks formed in lakes

are much more likely to contain the remains of land animals, than those that were formed in seas, and they are more likely to be in a complete state of preservation. Within the last century, five or six small lakes in Scotland, which had been artificially drained, yielded the remains of several hundred skeletons of stags, oxen, boars, horses, sheep, dogs, hares, foxes, and wolves. There are two ways in which these animals may have met with a watery grave. In the first place, they may have got mired on going into the water, or in trying to land on the other side, after swimming across. Any one who knows Scotch lakes will be familiar with the fact that their margins are often most treacherous ground for bathers. The writer has more than once found it necessary to be very cautious on wading into a lake while fishing, or in search of plants. Secondly, when such lakes are frozen over in winter, the ice is often very treacherous in consequence of numerous springs; and animals attempting to cross may be easily drowned. No remains of birds were discovered in these lakes, in spite of the fact that, until drained, they were largely frequented by water-fowl. But it must be remembered that birds are protected by their powers of flight from perishing in such ways as other animals frequently do. And, even should they die on the water, their bodies are not likely to be submerged; for, being light and feathery, they do not sink, but continue floating until the body rots away, or is devoured by some creature such as a hungry pike. For these reasons the remains of birds are unfortunately very rare in the stratified rocks; and hence our knowledge of the bird life of former ages is slight.

THE IMPERFECTION OF THE RECORD.

A very little consideration will serve to convince us that the record which Nature has kept in the stratified rocks is an incomplete one. There are many reasons why it must be so. It is

not to be expected that these rocks should contain anything
like a complete collection of the remains of the various tribes
of plants and animals that from time to time have flourished in
seas, lakes, and estuaries, or on islands and continents of the
world. In endeavouring to trace the course of life on the globe
at successive periods, we are continually met by want of evidence
due to the " imperfection of the record "—to use Darwin's phrase.
The reasons are not far to.seek. The preservation of organic
remains, or even of impressions thereof, in sedimentary strata is,
to some extent, a matter of chance. It is obvious that no wholly
soft creature, such as a jelly-fish, can be preserved ; although on
some strata they have left impressions telling of their existence
at a very early period.

A creature, to become fossilised, must possess some hard part,
such as a shell, *e.g.* an oyster (fossil oysters abound in some
strata) ; or a hard chitinous covering, like that of the shrimp, or
the trilobites of Silurian times ; or a skeleton, such as all the
backboned (vertebrate) animals possess.

But even creatures that had skeletons have not by any means
always been preserved. Bones, when left on the bottom of the
sea, where no sediment, or very little, is forming, will decay, and
so disappear altogether. As Darwin points out, we are in error
in supposing that over the greater part of the ocean-bed of the
present day sediment is deposited fast enough to seal up organic
remains before they can decay. Over a large part of the ocean-
bed such cannot be the case ; and this conclusion has, of late
years, been confirmed by the observations made during the
fruitful voyage of .H.M.S. *Challenger* in the Atlantic and Pacific
Oceans.

Again, even in shallower parts of the old seas, where sand or
mud was once deposited, fossilisation was somewhat accidental ; for
some materials, being porous, allow of the percolation of water,
and in this way shells, bones, etc., have been dissolved and lost.
Thus sandstone strata are always barren in fossils compared to

shales and limestones, which are much less pervious. To take examples from our own country, the New Red Sandstone of the south-west of England, the midland counties, Cheshire, and other parts contains very few fossils indeed, while the clays and lime-stones of the succeeding Lias period abound in organic remains of all sorts. Even insects have left delicate impressions of their wings and bodies! while shells, corals, encrinites, fish-teeth, and bones of saurians are found in great numbers.

Again, it must be borne in mind that the series of stratified rocks known to geologists is not complete or unbroken. They have been well compared to the leaves of a book on history, of which whole chapters and many separate pages have been torn out. These gaps, or " breaks," are due to what is called " denu-dation ;" that is to say, a great many rocks, after having been slowly deposited in water, have been upraised to form dry land, and then, being subjected for ages to the destroying action of "rain and rivers," or the waves of the sea, have been largely destroyed. Such rocks, in the language of geology, have been . " denuded ;" that is, stripped off, so that the underlying rocks are left bare.

But the process of rock-making does not go on continuously in any one area. Sedimentary strata have been formed in slowly sinking areas. But, if subsidence ceases, and the downward movement becomes an upward one, then the bed of the sea is converted into dry land, and the geological record is broken ; for aqueous strata do not form on dry land. Blown sands and terrestrial lava-flows are exceptions ; but such accumulations are very small and insignificant, and may therefore be neglected, especially as they contain no fossils.

In this way, as well as by the process of " denudation" already alluded to, breaks occur ; and these breaks often represent long intervals of time. There are several such gaps in the British series of stratified rocks ; and it is partly by means of these breaks, during which important geographical and other changes took place, that sedimentary rocks have been classified and arranged in groups representing geological periods. Thus, the Cainozoic, or Tertiary, rocks of the Thames' basin are separated by a long "break" from those of the preceding Cretaceous period. During that interval great changes in animal life took place, whereby, in the course of evolution, new types appeared on the scene. (See Table of Strata, Appendix I.)

Another cause interfering with the record is to be found in those important internal changes that have taken place in stratified rocks—often over large areas—which may be ascribed to the influence of heat and pressure combined. This process of change, whereby soft deposits have been altered or "metamorphosed" into hard crystalline rocks, is known as "metamorphism." Metamorphic rocks have lost not only their original structure and appearance, but also their included organic remains, or fossils. Thus, when a soft limestone has been converted by these means into crystalline statuary marble, any fossils it may once have contained have been destroyed. It is true that this applies more to older and lower deposits,—for the lowest are the oldest—but there can be no doubt that valuable records of the forms of life which peopled the world in former periods have been lost by this means.

And lastly, it must ever be borne in mind that, as yet, our knowledge of the stratified rocks of the earth's crust is very limited. In course of time, no doubt, this deficiency will be to a great extent made good ; but it will take a long time. Already, within the last thirty years, the labours of zealous geologists in the colonies and in various countries have added largely to our knowledge of the geological record. Still, only a small portion of the earth's surface has at present been explored ; and doubtless one may look forward to future discoveries of extinct forms of animal and plant life as wonderful and strange as those that have been of late years unearthed in the "far West," in Africa, and India. The Siwalik Hills of Northern India offer a rich harvest of fossils to future explorers. Already, one remarkable and large horned quadruped has come from this region ; and it is known that other valuable treasures are sealed up within these hills, only awaiting the "open sesame" of some enterprising explorer to bring them to light.

As previously pointed out, deposits formed in lakes are the most promising field for geologists in search of the remains of old terrestrial quadrupeds and reptiles ; but, unfortunately, such deposits are rare.

It is very much to be regretted that the carelessness and in-difference of ignorant workmen in quarries, clay-pits, and railway cuttings have sometimes been the cause of valuable fossils being broken up, and so lost for ever. Unless they are accustomed to the visits of fossil-collectors who will pay them liberally for their finds, the men will not take the trouble to preserve any bones they may come across in the course of their work. (An example of this negligence will be found on p. 95.) But when once they realise that such finds have what political economists call an "exchange value," or, in other words, can be turned into money, it is astonishing what zealous guardians of Nature's treasures they become ! For this reason collectors often find what Professor Bonney calls the "silver hammer"—in other words, cash—more effective than the iron implement they carry with them.

CHAPTER II.

SEA-SCORPIONS.

"And some rin up the hill and down dale, knapping the chucky slanes to pieces wi' hammers like sae many road-makers run daft. They say 'tis to see how the warld was made."—*St. Ronan's Well.*

OUR first group of monsters is taken from a tribe of armed warriors that lived in the seas of a very ancient period in the world's history. Like the crabs and lobsters inhabiting the coasts of Britain, they possessed a coat of armour, and jointed bodies, supplied with limbs for crawling, swimming, or seizing their prey. They were giants in their day, far eclipsing in size any of their relations that have lived on to the present time. Some of them, such as the Pterygotus (Fig. 1, p. 26), attained a length of nearly six feet. They belonged to the humbler ranks of life, and, if now living, would without doubt be assigned, by fishmongers ignorant of natural history, to that vague category of "shell-fish" in which they include crabs, lobsters, mussels, etc.

These lobster-like creatures, though claiming no relationship with the higher ranks of animals, may well engage our attention, not only for their great size, but also for their strange build.

There are no living creatures quite like them. Certainly they are not true lobsters, and yet we may consider them to be first or second cousins of those ten-footed crustaceans[1] of the present

[1] Crustaceans are a class of jointed creatures (articulate animals), possessing a hard shell or crust (Lat. *crusta*), which they cast periodically. They all breathe by gills.

SEA-SCORPIONS.

Pterygotus anglicus.
Length 6 feet.

Eurypterus.

Stylonurus.

PLATE I.

day—lobsters, crabs, and shrimps, so welcome on the tables of both rich and poor. Some naturalists say that their nearest relations at the present day are the king-crabs inhabiting the China seas and the east coast of North America; and there certainly are some points of resemblance between them. Others say that they are related to scorpions, and for this reason we call them Sea-scorpions. (See Plate I.)

The first feature we notice in these creatures is the way in which their bodies and limbs are divided into rings or joints. This fact tells us that they belong to that great division of animals called "Articulates," of which crabs, lobsters, spiders, centipedes, and insects are examples. The celebrated Linnæus called them *all* insects, because their bodies are in this way cut into divisions.[1] But this arrangement has since been abandoned. However, they are all built upon this simple plan, their bodies being like a series of rings, to which are attached paired appendages or limbs, also composed of rings, some longer and some shorter. Now, there must be something very fitting and appropriate in this arrangement, for the creatures that are thus built up are far more numerous than any other group of animals. They must be particularly well qualified to fight the battle of life; for like a victorious army they have taken the world by storm, and still remain in possession. We find them everywhere—in seas, rivers, and lakes; in fields and forests; in the soil, and in all sorts of nooks and crannies; in the air, and even upon or inside the bodies of other animals. Some of them, such as ants, bees, and wasps, show an intelligence that is simply marvellous, and have acquired social habits which excite our admiration.

Articulate animals are a very ancient race, as well as a flourishing one, for the oldest rocks containing undoubted fossils—namely, certain slates found in Wales and the Lake District—tell us of a time when shallow seas swarmed with little articulate animals known as *trilobites*. They were in appearance something like

[1] Lat. *in*, into, and *secta*, cut.

wood-lice of the present day; and the record of the rocks tells us plainly that creatures built upon this plan have flourished ever since. We mention this because they are related to the king-crabs of the present day, and therefore to the huge old-fashioned sea-scorpions we are now considering.

The best-known and largest of these creatures is represented in Fig. 1. It has received the name *Pterygotus* (or wing-eared) from certain fanciful resemblances pointed out by the quarrymen.

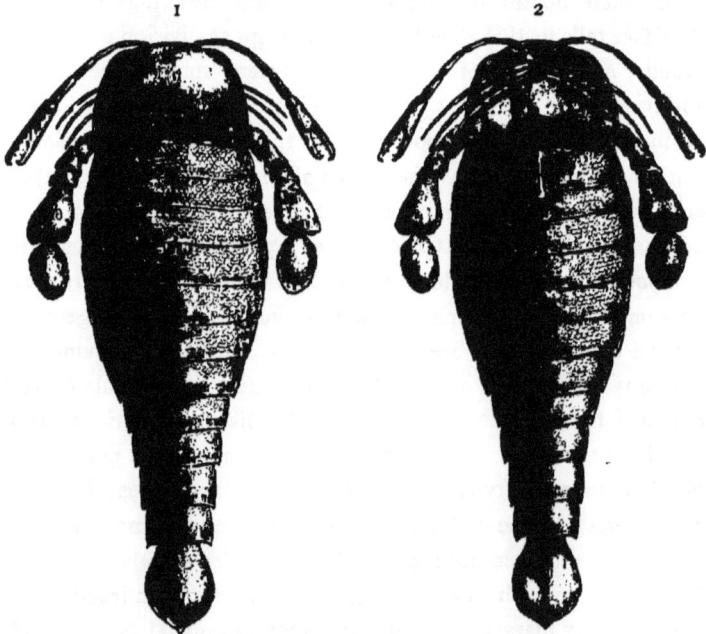

FIG. 1.—*Pterygotus anglicus.* (After Woodward.)
1. Upper side. 2. Under side.

It was first discovered, along with others of its kind, by Hugh Miller, at Carmylie in Forfarshire, in a certain part of the Old Red Sandstone (see Table of Strata, Appendix I.) known as the Arbroath paving-stone. The quarrymen, in the course of their work, came upon and dug out large pieces of the fossilised remains of this creature. Its hard coat of jointed armour bore on its surface curious wavy markings that suggested to their minds the sculptured feathers on the wings of cherubs—of all subjects of the chisel the most common. Hence they christened these remains "Seraphim." They did not succeed in getting complete specimens that could be pieced together; and the part to which this fanciful name was given turned out to be part of the under side below the mouth. It was composed of several large plates, two of which are not unlike the wings of a cherub in shape. Hugh Miller says in his classic work, *The Old Red Sandstone*—"the form altogether, from its wing-like appearance, its feathery markings, and its angular points, will suggest to the reader the origin of the name given it by Forfarshire workmen."

A correct restoration, in proportion to the fragments found in the Lower Old Red Sandstone, would give a creature measuring nearly six feet in length, and more than a foot across. *Pterygotus anglicus* may therefore be justly considered a monster crustacean.

The illustrious Cuvier, who, in the eighteenth century founded the science of comparative anatomy (see p. 5), astonished the scientific world by his bold interpretations of fossil bones. From a few broken fragments of bone he could restore the skeleton of an entire animal, and determine its habits and mode of living. When other wise men were unable to read the writing of Nature on the walls of her museum—in the shape of fossil bones—he came forward, like a second Daniel, to interpret the signs, and so instructed us how to restore the world's lost creations. Hugh Miller submitted the fragments found at Balruddery to the celebrated naturalist Agassiz, a pupil of Cuvier, who had written a famous work on fossil fishes; and he says that he was much struck with the skill displayed by him in piecing together the fragments of the huge Pterygotus. "Agassiz glanced over the collection. One specimen especially caught his attention—an elegantly symmetrical one. His eye brightened as he contemplated it. ' I will tell you,' he said, turning to the company—' I

will tell you what these are—the remains of a huge lobster.' He
arranged the specimens in the group before him with as much
ease as I have seen a young girl arranging the pieces of ivory in
an Indian puzzle. There is a homage due to supereminent genius,
which Nature spontaneously pays when there are no low feelings
of jealousy or envy to interfere with her operations; and the
reader may well believe that it was willingly rendered on this
occasion to the genius of Agassiz." Agassiz himself, previous to
this, had considered such fragments as he had seen to be the
remains of fishes. As we have said before, this creature was *not*
a true lobster; but Agassiz, when he expressed the opinion just
quoted, was not far off the mark, and did great service in showing
it to be a crustacean. There were no lobsters or scorpions at
that early period of the world's history, and this creature, with its
long "jaw-feet" and powerful tail, was a near approach to a
king-crab on the one hand and scorpion on the other. If living
now, it would no doubt command a high price at Billingsgate;
but, then, it would be a dangerous thing to handle when alive,
and might be more troublesome to catch than our crabs or
lobsters.

The front part of its body was entirely enveloped in a kind
of shield, called a carapace, bearing near the centre minute
eyes, which probably were useless, and at the corners two large
compound eyes, made up of numerous little lenses, such as we
see in the eye of a dragon-fly. This is clearly proved by certain
well-preserved specimens. There are five pairs of appendages,
all attached under or near the head. Behind the head follow
thirteen rings, or segments, the last of which forms the tail,
two at least of these bore gills for breathing. All but two of
them, below the mouth, must have been beautifully articulated, so
as to allow them to move freely, as we see in the lobster of the
present day. But look at that lowest and largest pair of appendages,
the end joints of which are flattened out, and you will see that
they must have been a powerful oar-like apparatus for swimming

forwards. We can fancy this creature propelling itself much in the same way as a "water-beetle" rows itself through the water in a pond. In all other crustaceans the antennæ are used for feeling about, but in the Pterygotus they are used as claws for seizing the prey.

In general external appearance, this huge Pterygotus greatly reminds us of a tiny fresh-water crustacean, known as Cyclops—because it has only one eye, like the giant in Homer's *Odyssey*. This little creature, which is only $\frac{1}{16}$ inch in length, is an inhabitant of ponds. From its large eyes, powerful oar-like limbs, or appendages, and from the general form of its body, Dr. Henry Woodward (the author of a learned monograph on these creatures) concludes that the Pterygotus was a very active animal; and the reader will easily gather from its pair of antennæ, converted at their extremities into nippers, and from the nature of its "jaw-feet," that the creature was a hungry and predaceous monster, seizing everything eatable that came in its way. The whole family to which it belongs—including Pterygotus, Eurypterus, Slimonia, Stylonurus, and others—seems to have been fitted for rather rapid motion, if we may judge from the long tapering and well-articulated body. In two forms (Pterygotus and Slimonia) the tail-flap probably served both as a powerful propeller, and as a rudder for directing the creature's course; but others, such as Eurypterus and Stylonurus, had long sword-like tails, which may have assisted them to burrow into the sand, in the same way that king-crabs do. Eurypterus remipes is shown in Fig. 2.

It has been stated above that our sea-scorpions are related to the king-crabs. Now, this creature, it is well known, burrows into the mud and sand at the bottom of the sea. This it does by shoving its broad sharp-edged head-shield downwards, working rapidly at the same time with its hinder feet, or appendages, and by pushing with the long spike that forms a kind of tail. It will thus sink deeper and deeper until nothing can be seen of its body, and only the eyes peep out of the mud. It will crawl and wander about by night, but remains hidden by day. Some of them are of large size, and occasionally measure two feet in length. They possess six pairs of well-formed feet, the joints of which, near the body, are armed with teeth and spines, and serve

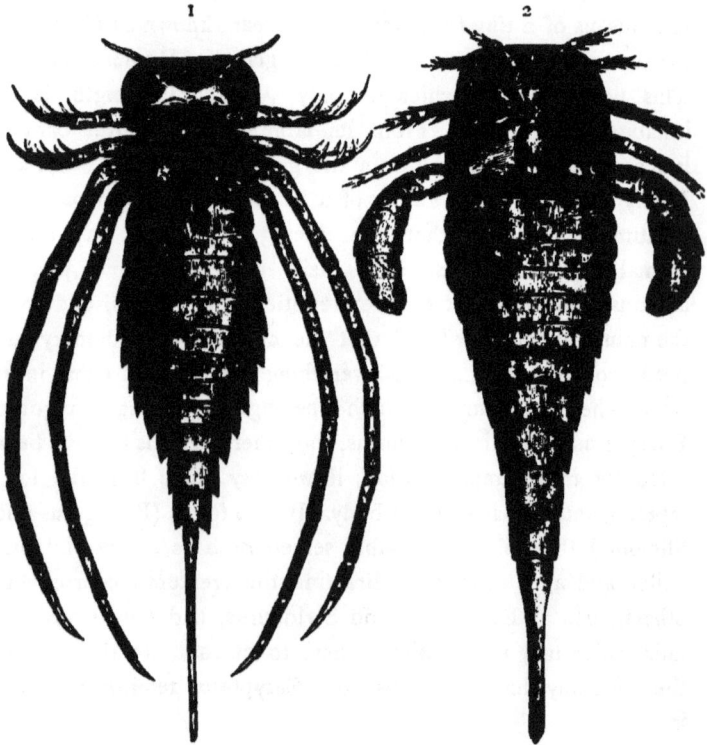

FIG. 2.—*Silurian merostomata.*

1. *Stylonurus.* 2. *Eurypterus.* (After Woodward.)

the purpose of jaws, being used to masticate the food and force it into the mouth, which is situated between them.

Now, this fact is of great importance; for it helps us to understand the use of the four pairs of "jaw-feet" in our Sea-scorpions. What curious animals they must have been, using the same limbs for walking, holding their prey, and eating ! Look at the broad plates at the base of the oar-like limbs, or appendages, with their tooth-like edges. These are the plates found by Hugh Miller's quarrymen, and compared by them to the wings of seraphim. You will easily perceive that by a backward and forward movement, they would perform the office of teeth and jaws, while the long antennæ with their nippers— helped by the other and smaller appendages—held the unfortunate victim in a relentless grasp. And even these smaller limbs, you will see from the figure, had their first joints, near the mouth, provided with toothed edges like a saw.

With regard to the habits of Sea-scorpions, it would not be altogether safe to conclude that, because in so many ways they resembled king-crabs, they therefore had the same habit of burrowing into the soft muddy or sandy bed of the sea, as some authorities have supposed. Seeing that there is a difference of opinion on this subject, the author consulted Dr. Woodward on the question, and he said he thought it unlikely, seeing that, in some of them, such as the Pterygotus, the eyes are placed on the margin of the head-shield ; for it would hardly care to rub its eyes with sand. Whether it chose at times to bury its long body in the sand by a process of wriggling backwards, as certain modern crustaceans do, we may consider to be an open question.

If only Sea-scorpions had not unfortunately died out, how interesting it would be to watch them alive, and to see exactly what use they would make of their long bodies, tail-flaps, and tail-spikes ! Were they nocturnal in their habits, wandering about by night, and taking their rest by day? Such questions, we fear, can never be answered. But their large eyes would have been able to collect a great deal of light when the moon and stars feebly illumined the shallower waters of the seas of Old Red Sandstone times ; and so there is nothing to contradict the idea.

Now, it is an interesting fact that young crabs, soon after they

are hatched, have long bodies somewhat similar to those of our
Sea-scorpions, with a head-shield under which are their jaw-feet, and
then a number of free body-rings without any appendages.   These
end in a spiked tail.   As the crab grows older, he ceases to be
a free-swimming animal—for which kind of life his long body
is well suited,—tucks up his long tail, and takes to crawling
instead.   Thus his body is rendered more compact and handy
for the life he is going to lead.   Lobsters, on the other hand, can
swim gently forwards, or dart rapidly backwards.   Thus we see
that the ten-footed crustaceans of the present day are divided into
two groups—the long-tailed and free-swimming forms, such as
lobsters, shrimps, and cray-fishes; and the short-tailed crawling
forms, namely, the crabs.   Now, in the same way, Pterygotus and
its allies were long-tailed forms, while the king-crabs are short-
tailed forms.   So were the trilobites of old.   Hence we learn
that, ages and ages ago, before the days of crabs and lobsters,
there were long-tailed and short-tailed forms of crustaceans, just
as there are now, only they did not possess true walking legs.
They belonged to quite a different order, called "thigh-mouthed"
crustaceans, Merostomata, because their legs are all placed near the
mouth ; and, as we have already learned, were used for feeding
as well as for purposes of locomotion.

Now, one of the many points of interest in Pterygotus and its
allies is that they somewhat resemble the crab in its young or
larval state.   To a modern naturalist, this fact is important as
showing that crustacean forms of life have advanced since the
days of the sea-scorpions.

Their resemblance to land-scorpions is so close that, if it were
not for the important fact that scorpions breathe *air* instead of
water, and for this purpose are provided with air-tubes (or
trachea) such as all insects have, they would certainly be removed
bodily out of the crustacean class, and put into that in which
scorpions and spiders are placed, viz. the Arachnida.   But, in
spite of this important difference, there are some naturalists in

favour of such a change. It will thus be seen that our name
Sea-scorpions is quite permissible.

Hugh Miller described some curious little round bodies found
with the remains of the Pterygotus, which it was thought were
the eggs of these creatures!

Finally, these extinct crustaceans flourished in those ages of
the world's history known as the Silurian and the Old Red Sandstone periods. As far as we know, they did not survive beyond
the succeeding period, known as the Carboniferous.[1]

[1] The student should consult Dr. Henry Woodward's valuable *Monograph
of the British Merostomata* (Palæontographical Society), to which the writer is
much indebted. With regard to the representation of *Pterygotus anglicus* in
Plate I., it has been pointed out by Dr. Woodward that the creature was
unable to bend its body into such a position as is shown there. As in a
modern lobster, or shrimp, there were certain overlapping plates in the rings,
or segments, of the body, which prevented movement from side to side, and
only allowed of a vertical movement.

D

CHAPTER III.

THE GREAT FISH-LIZARDS.

" Berossus, the Chaldæan saith : A time was when the universe was darkness and water, wherein certain animals of frightful and compound forms were
generated.   There were serpents and other creatures with the mixed shapes of
one another, of which pictures are kept in the temple of Belus at Babylon."—
The Archaic Genesis.

VISITORS to Sydenham, who have wandered about the spacious
gardens so skilfully laid out by the late Sir Joseph Paxton, will
be familiar with the great models of extinct animals on the
"geological island."   These were designed and executed by that
clever artist, Mr. Waterhouse Hawkins, who made praiseworthy
efforts to picture to our eyes some of the world's lost creations, as
restored by the genius of Sir Richard Owen and other famous
naturalists.   His drawings of extinct animals may yet be seen
hanging on the walls of some of our provincial museums ; and
doubtless others still linger among the natural history collections
of schools and colleges.

Lazily basking in the sun, when it condescends to shine, and
resting his clumsy carcase on the ground that forms the shore
near the said geological island at Sydenham, may be seen the old
fish-lizard, or Ichthyosaurus, that forms the subject of the present
chapter.   He looks awkward on land, as if longing to get into his
native element once more, and cleave its waters with his powerful tail-fin.   His "flippers" seem too weak to enable him to crawl
on land.   Moreover, the most recent discoveries of Dr. Fraas

lead us to conclude that the Ichthyosaur never ventured to leave the "briny ocean" to bask upon the land.

This great uncouth beast presents some curious anomalies in his constitution, being planned on different lines to anything now living, and presenting, as so many other extinct animals do, a mixture, or fusion, of types that greatly puzzled the learned men of the time when his remains were first brought to light, after their long entombment in the Lias rocks forming the cliffs on the coast of Dorset. Some have christened him a "sea-dragon," and such indeed he may be considered. But the name Ichthyosaurus, given above, has received the sanction of high authority, and, moreover, serves to remind us of the fact that, although in many respects a lizard, he yet retains in his bony framework the traces of a remote fishy ancestry. So we will call him a fish-lizard.

We remember in our young days the amiable endeavours of Mr. "Peter Parley" to introduce us to the wonders of creation; and his account of the Ichthyosaurus particularly impressed itself on our youthful imagination. How surprised that inestimable instructor of youth would be could he now see the still more wonderful remains that have been brought to light from Europe, Asia, Africa, and America!

The curious quotation given at the head of the present chapter refers to a widespread belief, prevalent among the highly civilised nations of antiquity, that the world was once inhabited by dragons, or other monsters "of mixed shape" and characters. To the student of ancient history traces of this curious belief will be familiar. Sir Charles Lyell refers to such a belief when he says, in his *Principles of Geology*, "The Egyptians, it is true, had taught, and the Stoics had repeated, that the earth had once given birth to some monstrous animals that existed no longer." It may be surprising to some, but it is undoubtedly the fact, that modern scientific truths were partly anticipated by the civilised nations of long ago. Take the ideas of the ancients as interpreted from the records of Egypt, Chaldæa,

India, and China; and you will find that our discoveries in geology, astronomy, and ethnology go far to prove that the traditions of these ancient peoples, however derived, after making due allowance for Oriental allegory and poetic hyperbole, are not far from the truth. To the Babylonian tradition of the monstrous forms of life at first created we have already alluded; but in other fields of discovery we find the same foreshadowing of discoveries made in our own day. Take the vast cycles of Egyptian tradition, wherein the stars returned to their places after a circle of constant change, only to start again on their unwearied round; the atomic theory of Lucretius, now expanded and incorporated into modern chemistry; or the philosopher's pregnant saying—*Omne vivum ex ovo* (" Every living thing comes from an egg "). These and other examples might be cited to show how true the old saying is, " There is nothing new under the sun." In the writings of ancient authors may be found singular notices of bones and skeletons found in "the bowels of the earth," which are referred to an imaginary era of long ago, when giants of huge dimensions walked this earth. One is inclined sometimes to wonder whether the old fables of griffins and horrid dragons may not be to some extent based upon the occasional discovery, in former times, of fossil bones, such as evidently belonged to animals the like of which are not to be seen nowadays. (See chaps. xiii. and xiv.)

The illustrious Cuvier, in his day, considered the fish-lizard to be one of the most heteroclite and monstrous animals ever discovered. He said of this creature that it possessed the snout of a dolphin, the teeth of a crocodile, the head and breast-bone of a lizard, the paddles of a whale or dolphin, and the vertebræ of a fish! No wonder that naturalists and palæontologists, whose realm is the natural history of the past, were obliged to make a new division, or order, of reptiles to accommodate the fish-lizard. It is obvious that a creature with such very "mixed" relationships would be out of place in any of the four orders into which living reptiles, as represented by turtles, snakes, lizards, and crocodiles are divided. Here is what Professor Blackie says of the Ichthyo-saurus—

> " Behold, a strange monster our wonder engages !
> If dolphin or lizard your wit may defy.
> Some thirty feet long, on the shore of Lyme-Regis,
> With a saw for a jaw, and a big staring eye.
> A fish or a lizard ? An ichthyosaurus,
> With a big goggle eye, and a very small brain,
> And paddles like mill-wheels in chattering chorus,
> Smiting tremendous the dread-sounding main."

A glance at our restoration, Plate II., will show that the fish-lizard was a powerful monster, well endowed with the means of propelling itself rapidly through the water as it sought its living prey, to seize it within those cruel jaws. The long and powerful tail was its chief organ of propulsion ; but the paddles would also be useful for this purpose, as well as for guiding its course. The pointed head and generally tapering body suggests . a capability of rapid movement through the water ; and since we know for certain that it fed on fishes, this conclusion is con-firmed, for fishes are not easily caught now, and most probably were not easily caught ages ago.

The personal history of the fish-lizard, merely as a fossil or "remain," is interesting; so much so, that we may perhaps be allowed to relate the circumstances of his *début* before the scientific world, in the days of the ever-illustrious Cuvier, to whom we have already alluded. But England had its share of illustrious men, too, though lesser lights compared to the founder of comparative anatomy,—such as Sir Richard Owen, on whom the mantle of his friend Cuvier has fallen; Conybeare, De la Beche, and Dean Buckland.

These scientific men, aided by the untiring labours of many enthusiastic collectors of organic remains, have been the means of solving the riddle of the fish-lizard, and of introducing him to the public. By this time there is, perhaps, no creature among the host of Antediluvian types better known than this reptile.

The remains of fish-lizards have attracted the attention of collectors and describers of fossils for nearly two centuries past. The vertebræ, or "cup-bones," as they are often called, of which the spinal column was composed, were figured by Scheüchzer, in an old work entitled *Querelæ Piscium;* and, at that time, they were supposed to be the vertebræ of fishes. In the year 1814 Sir Everard Home described the fossil remains of this creature, in a paper read before the Royal Society, and published in their *Philosophical Transactions.* This fossil was first discovered in the Lias strata of the Dorsetshire coast. Other papers followed till the year 1820. We are chiefly indebted to De la Beche and Conybeare for pointing out and illustrating the nature of the fish-lizard ; and that at a time when the materials for so doing were far more scanty than they are now. Mr. Charles König, Mr. Thomas Hawkins, Dean Buckland, Sir Philip Egerton, and Professor Owen have all helped to throw light on the structure and habits of these old tyrants of the seas of that age, which is known as the Jurassic period. They lived on, however, to the succeeding or Cretaceous period, during which our English chalk was forming; but the Liassic age was the one in which they flourished most abundantly, and developed the greatest variety.

In the year 1814 a few bones were found on the Dorsetshire coast between Charmouth and Lyme-Regis, and added to the collection of Bullock.. They came from the Lias cliffs, under-mined by the encroaching sea. Sir Everard's attention being attracted to them, he published the notices already referred to. The analogy of some of the bones to those of a crocodile, induced Mr. König, of the British Museum, to believe the animal to have been a saurian, or lizard ; but the vertebræ, and also the position of certain openings in the skull, indicated some remote affinity with fishes, but this must not be pressed too far. The choice of a name, therefore, involved much difficulty ; and at length he decided to call it the *Ichthyosaurus,* or fish-lizard. Mr. Johnson, of Bristol, who had collected for many years in that neighbourhood, found out some valuable particulars about these remains. The conclusions of Dean Buckland, then Professor of Geology at Oxford, led Sir Everard to abandon many of his former conclusions. The labours of the learned men of the day were greatly assisted by the exertions of Miss Anning, an enthusiastic collector of fossils. This lady, devoting herself to science, explored the frowning and precipitous cliffs in the neighbourhood of Lyme-Regis, when the furious spring-tide combined with the tempest to overthrow them, and rescued from destruction by the sea, sometimes at the peril of her life, the few specimens which originated all the facts and speculations of those persons whose names will ever be remembered with gratitude by geologists.

Probably our readers are already more or less familiar with the drawings of the fossilised remains of Ichthyosauri to be seen in almost every text-book of geology. (Fig. 3 is from Owen's *British Fossil Reptiles.*) But we recommend all who take an interest in the

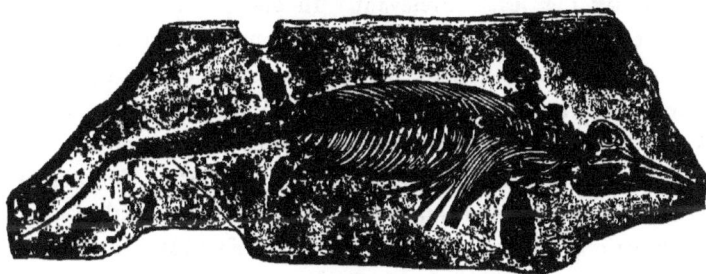

FIG. 3.—*Ichthyosaurus intermedius.*

world's lost creations to pay a visit to the great Natural History Museum, at South Kensington. The fossil reptile gallery contains a magnificent series of Ichthyosauri, about thirty in number. Of these a large number were obtained through the exertions of the late Mr. T. Hawkins, a Somersetshire gentleman, who was a most ardent collector of fossil reptiles, and who devoted himself with great enthusiasm and unsparing energy to the acquisition of a truly splendid collection of these most interesting relics of the past. Nearly sixty years ago he arranged for the purchase of his treasures by the authorities of the British Museum, and thus his collection became the property of the nation.

His specimens were figured and described by him in two large folio volumes. The first was published in 1834, under the title, *Memoirs of the Ichthyosauri and Plesiosauri;* his second, with the same plates, in 1842, under the quaint title of *The Book of the Great Sea-Dragons.* The large lithographic drawings of his fine specimens were beautifully executed by Scharf and O'Neil. The plates are the only really valuable part of these two curious and ill-written books. His descriptions are not of much value, and his pages are encumbered with a vast amount of extraneous matter. The author is immensely proud of his collection, and his vanity is conspicuous throughout. Instead of confining himself to descriptions of what he found, and how he found them, he continually wanders into all sorts of subjects that are, to say the least, irrelevant. In one place he introduces ancient history and mythology; in another, Old Testament chronology; in another, the unbelieving spirit of the age; and here and there indulges in vague unphilosophical speculations. Altogether his two volumes are a curious mixture of bigotry, conceit, and unrestrained fancy, and they afforded to the present writer no small amusement. One rises from the perusal of such men's writings with a strong sense of the contrast between the humble and patient spirit in which our great men of to-day, such as Professor Owen, study nature and record their observations, and the vague, conceited outpourings of some old-fashioned writers.

Mr. Hawkins tells us that his youthful attention was directed to the Lias quarries, near Edgarly, in Somersetshire, in consequence of some strange reports. It was said that the bones of giants and infants had, at distant intervals, been found in them. These quarries he visited, and, by offers of generous payment, induced the workmen to keep for him all the remains they might

FISH-LIZARDS.

Fishes, *Dapedius*, etc.

Ichthyosaurus tenuirostris.
A smaller species.

Ichthyosaurus communis.
Length about 22 feet.

PLATE II.

find. In this way he finally obtained the co-operation of all the quarrymen in the county.

Mr. Hawkins thus expresses his delight on obtaining an Ichthyosaurus which was pointed out to him by Miss Anning, near the church at Lyme-Regis, in the year 1832 : "Who can describe my transport at the sight of the colossus? My eyes the first which beheld it! Who shall ever see them lit up with the same unmitigated enthusiasm again? And I verily believe that the uncultivated bosoms of the working men were seized with the same contagious feeling; for they and the surrounding spectators waved their hats to an ' Hurra !' that made hill and mossy dell echoing ring."

This specimen, however, got sadly broken in its fall from the cliff; but in time he put all the pieces together again. Speaking of his own collection, he says, " This stupendous treasure was gathered by me from every part of England; arranged, and its multifarious features elaborated from the hard limestone by my own hands. A tyro in collecting at the age of twelve years, I then boasted of all the antiquities that were come-at-able in my neighbourhood, but, finding that everybody beat my cabinet of coins, I addressed myself to worm-eaten books, and last to fossils." Before he was twenty years of age. he had obtained a very fine collection of organic remains.

When, however, he complains of the Philistine dulness and stupidity of quarrymen, who often, in their ignorance, break up finds of almost priceless value, we can fully sympathize.

In general contour the body of the fish-lizard was long and tapering, like that of a whale (see Plate II.). It probably showed no distinct neck. The long tail was its chief organ of propulsion. We notice two pairs of fins, or paddles ; one on the fore part of the body, the other on the hinder part, like the pectoral and abdominal fins of a fish. The skin was scaleless and smooth, or slightly wrinkled, like that of a whale. No traces of scales have ever been found ; and if such had existed, they would certainly have been preserved, since those of fishes and crocodiles of the Jurassic period have been found in considerable number and variety. It is therefore safe to conclude that such were absent in this case. In the Lias strata, at least, the specimens are often preserved with most wonderful completeness (see p. 47).

The long and pointed jaws are a striking feature of these animals. The eyes were very large and powerful, and specially adapted, as we shall see presently, to the conditions of their life.

It might, perhaps, be asked whether the fish-lizards breathed, like fishes, by means of gills. That question can easily be answered; for if they had possessed gills for taking in water and breathing the air dissolved therein, they would reveal the fact by showing a bony framework for the support of gills, such as are to be found in all fishes. These structures, known as "branchial arches," are absent; therefore the fish-lizards possessed lungs, and breathed air like reptiles of the present day. Their skulls show where the nostrils were situated; namely, near the eyes, and not at the end of the upper jaw-bone. There are also passages in the skull leading from the nostrils to the palate, along which currents of air passed on their way to the lungs. Being air-breathers, they would be compelled occasionally to seek the surface of the sea, in order to obtain a fresh supply of the life-giving element—oxygen; but, being cold-blooded and with a small brain, needing a much less supply of oxygen for its work, the fish-lizards had, like fishes, this advantage over whales, which are warm-blooded—that their stern-propeller, or tail-fin, could take the form best adapted for a swift, straight-forward course through the water.

In the whale tribe the tail-fin is horizontal; and this is so on account of their need, as large-brained, warm-blooded air-breathers, of speedy access to the atmospheric air. Were it otherwise, they would not have the means of rising with sufficient rapidity to the surface of the sea; for they have only one pair of fins. But the fish-lizards had two pairs of these appendages, and the hinder or pelvic pair no doubt were of great service in helping the creatures to come up to the surface when necessary.

Thus we see that the whale, with its one pair of paddles, has a tail specially planned with a view to rapid vertical movement through the water; while in the fish-lizards, who did not require to breathe so frequently, the tail-fin was planned with a view to swift and straight movement forward as they pursued their prey, and they were compensated by having bestowed upon them an extra pair

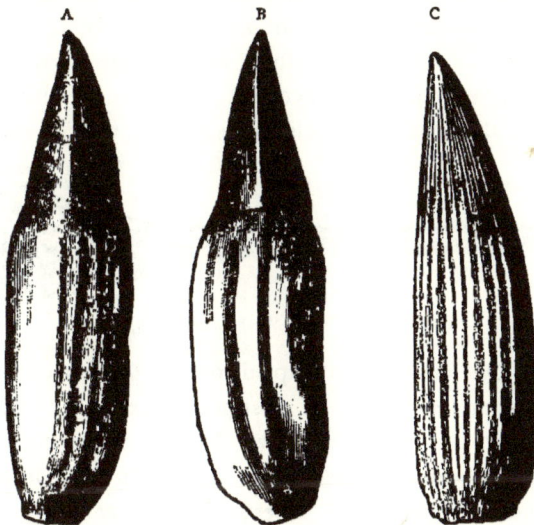

FIG. 4.—(A) Lateral and (B) profile views of a tooth of *Ichthyosaurus platyodon* (Conybeare), Lower Lias, Lyme Regis, Dorsetshire. (C) Tooth of *Ichthyosaurus communis* (Conybeare), Lower Lias, Lyme Regis, Dorset.

of paddles. Thus we learn how one part of an animal is related to and dependent upon another, and how they all work together with the greatest harmony for certain definite purposes (see p. 6).

These great marine predaceous reptiles literally swarmed in the seas of the Lias period, and no doubt devoured immense

shoals of the fishes of those times, whose numbers were thus
to some extent kept down.   There is clear proof of this in the
fossilised droppings—known as "coprolites,"—which show on
examination the broken and comminuted remains of the little
bony plates of ganoid fishes that we know were contemporaries
of these reptiles.   Probably young ones were sometimes devoured
too.

It was in the period of the Lias that fish-lizards attained to their
greatest development, both in numbers and variety; and the
strata of that period have preserved some interesting variations.
It will be sufficient here to point out two, namely, Ichthyosaurus
tenuirostris—an elegant little form, in which the jaws, instead of
being massive and strong, were long and slender like a bird's
beak; and also Ichthyosaurus latifrons (Fig. 5), with jaws still more

FIG. 5.—Skull of *Ichthyosaurus latifrons.*

birdlike.   Our artist has attempted to show the former variety in
our illustration (Plate II.).   A most perfect example of this pretty
little Ichthyosaur, from the Lower Lias of Street in Somerset,
has recently been presented to the National Collection at South
Kensington by Mr. Alfred Gillett, of Street, and may be seen
there.   In this group of fish-lizards the eyes are relatively larger,
and we should imagine that they were very quick in detecting
and catching their prey; their paddles also have larger bones.

There is a remarkably fine specimen at Burlington House, in
the rooms of the Geological Society, of an Ichthyosaurus' head,
which the writer found, on measuring, to be about five feet six
inches long.   A cast of this head is exhibited at South Kensington.
The largest of the specimens in the National Collection is twenty-
two feet long and eight feet across the expanded paddles; but it

is known that many attained much greater dimensions. Judging from detached heads and parts of skeletons, it is probable that some of them were between thirty and forty feet long. A specimen of Ichthyosaurus platyodon in the collection of the late Mr. Johnson, of Bristol, has an eye-cavity with a diameter of fourteen inches. This collection is now dispersed.

With regard to their habits, Sir Richard Owen concludes that they occasionally sought the shores, crawled on the strand, and basked in the sunshine. His reason for this conjecture (which, however, is not confirmed by Dr. Fraas's recent discoveries) is to be found in the bony structure connected with the fore-paddles, which is not to be found in any porpoise, dolphin, grampus, or whale, and for want of which these creatures are so helpless when left high and dry on the shore.[1] The structure in question is a strong bony arch, inverted and spanning across beneath the chest from one shoulder to the other. A fish-lizard, when so visiting the shore for sleep, or in the breeding season, would lie or crawl, prostrate, with its under side resting or dragging on the ground— somewhat after the manner of a turtle.

It is a curious fact that this bony arch resembles the same part in those singular and problematical mammals, the Echidna and the Platypus, or duck-mole.

The enormous magnitude and peculiar construction of the eye are highly interesting features. The expanded pupil must have allowed of the admittance of a large quantity of light, so that the creature possessed great powers of vision.

The organic remains associated with fish-lizards tell us that they inhabited waters of moderate depth, such as prevails near a coast-line or among coral islands. Moreover, an air-breathing creature would obviously be unable to live in "the depths of the

---

[1] It is, perhaps, hardly necessary to remark that whales are not fishes, but mammals which have undergone great change in order to adapt themselves to a marine life. Their hind limbs have practically vanished, only a rudiment of them being left.

sea ; " for it would take a long time to get to the surface for a fresh supply of air.

Perhaps no part of the skeleton is more interesting than the curious circular series of bony plates surrounding the iris and pupil of the eye.   The eyes of many fishes are defended by a bony covering consisting of two pieces ; but a circle of bony overlapping plates is now only found in the eyes of turtles, tortoises, lizards, and birds, and some alligators.   This elaborate apparatus must have been of some special use; the question is—What service or services did it perform?   Here, again, we find answers suggested by Owen and Buckland.   It would aid, they say, in protecting the eye-ball from the waves of the sea when the creature rose to the surface, as well as from the pressure of the water when it dived down to the bottom—for even at a slight depth pressure increases, as divers know.   But it appears that the ring of bony plates fulfilled a yet more important office, thereby enabling the fish-lizards to play admirably their part in the world in which they lived, and to succeed in the struggle of life; for even in those remote days there must have been, as now, a keen competition among all animals, so that the victory was to those that were best equipped.

Would it not be an advantage for them to have the power of seeing their finny prey whether near or far?   Certainly it would ; and so we are told that, by bringing the plates a little nearer together, and causing them to press gently on the eye-ball, so as to make the eye more convex—that is, bulging out—a nearer object would be the better discerned.   On the other hand, by relaxing this pressure, thus enlarging the aperture of the pupil and diminishing the convexity, a distant object would be focussed upon the retina.   In this manner some birds alter the focus of their eyes while swooping down on their prey.

What a wonderful arrangement ! We often hear of people having two pairs of spectacles—with lenses of different curvature—one for reading, and the other for seeing more distant objects than a book held in the hand. But here is a creature that possessed an apparatus far more simple and effective than that supplied by the optician ! Dr. Buckland, speaking of these "sclerotic plates," as they are called, says they show "that the enormous eye of which they formed the front was an optical instrument of varied and prodigious power, enabling the Ichthyosaurus to descry its prey in the obscurity of night and in the depths of the sea." But the last expression must be taken in a limited sense (see Fig. 6).

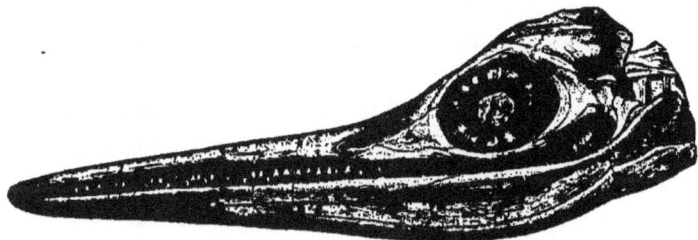

FIG. 6.—Head of *Ichthyosaurus platyodon*.

It might well be supposed that no record had been preserved from which we could learn anything about the nature of the skin of our fish-lizard ; but even this wish has been partly fulfilled, to the delight of all geologists. Certain specimens have been obtained, from the Lias of England and Germany, that show faithful impressions of the skin that covered the paddles. A specimen of this nature has lately been presented to the national treasure-house at South Kensington by Mr. Montague Brown. On the inner side of the paddle was a broad fin-like expansion, admirably adapted to obtain the full advantage of the stroke of the limb in swimming.[1]

Speaking of the limbs, it should be mentioned that the bones of each finger, instead of being elongated and limited in number to three in each of the five fingers, are polygonal in shape and

[1] Mr. Smith Woodward informs the writer that specimens have lately been found near Würtemberg, with evidence of a triangular fin on the back. - Plate II. has been redrawn for this edition, to make it more in harmony with Dr. Fraas's discoveries. (See Appendix V.)

arranged in as many as seven or eight rows, while those of each finger are exceedingly numerous. Thus the whole structure forms a kind of· bony pavement which must have been very supple. Such a limb would ·be one of the most efficient and powerful swimming organs known in the whole animal kingdom. In whales the fingers of the flippers are of the usual number, namely, five. Some species of fish-lizards had as many as over a hundred separate little bones in the fore-paddle.

Another question naturally suggests itself: Were they viviparous, or did they lay eggs like crocodiles? This question seems to have been answered in favour of the first supposition; and in the following interesting manner. It not infrequently happens that entire little skeletons of very small individuals are found under the ribs of large ones. They are invariably uninjured, and of the same species as the one that encloses them, and with the head pointing in one direction. Such specimens are most probably the fossilised remains of little fish-lizards, that were yet unborn when their mothers met with an untimely end (see p. 51). In some cases, however, they may be young ones that were swallowed. (See Appendix V.)

The jaws of these hungry formidable monsters were provided with a series of formidable teeth—sometimes over two hundred in number—inserted in a long groove, and not in distinct sockets, as in the case of crocodiles. In some cases, sixty or more have been found on each side of the upper and lower jaws, giving a total of over two hundred and forty teeth! The larger teeth may be two inches or more in length.

The jaws were admirably constructed on a plan that combined lightness, elasticity, and strength. Instead of consisting of one piece only, they show a union of plates of bone, as in recent crocodiles. These plates are strongest and most numerous just where the greatest strength was wanted, and thinner and fewer towards the extremities of the jaw. A crocodile, Sir Samuel Baker says, in his *Wild Beasts and their Ways*, can bite a man in two; and no doubt our fish-lizard would have been glad to perform the same feat ! But in his pre-Adamite days the opportunity did not present itself.

The spinal column, or backbone, with its generally concave vertebræ, must have been highly flexible, as is that of a fish, especially the long tail which the creature worked rapidly from side to side as it lashed the waters.

The hollows of these concave vertebræ must have been originally filled up with fluid forming an elastic bag, or capsule. To get a clearer idea of this, take a small portion of the backbone of a boiled cod, or other "bony" fish, and you will see on pulling it to pieces, the white, jelly-like substance that fills up the hollows between the vertebræ. In this way Nature provides a soft cushion between the joints, that allows of a certain amount of movement, while, at the same time, the column holds together. The backbone of a fish may not inaptly be compared to a railway train. Each of the carriages represents a vertebra, and the buffers act as cushions when the train is bent in running round a curve. After all, we must learn from Nature; and many of the greatest mechanical and engineering triumphs of to-day are based upon the methods used by Nature in the building up and equipment of vegetable and animal forms of life.

It may, perhaps, be inquired whether there is any evidence for the existence of a tail-fin, such as is shown in our illustration. To this it may be replied that the presence of such an appendage is as good as proved by a certain flattening of the vertebræ at the end of the tail, detected by Owen. The direction of this flattening is from side to side, and therefore the tail-fin must have been vertical, like that of a fish. In one specimen Sir Richard Owen has detected as many as 156 vertebræ to the whole body.

Our description of the fish-lizard has, we trust, been sufficient—

E

although not couched in the language used by men of science—to give a fair idea of its structure and habits.

In conclusion, a few words may be said about the ancestry and life-history of these ancient monsters. Palæontologists have good reason to believe that they were descended from some early form of land reptile. If so, they show that whales are not the first land animals that have gone back to the sea, from which so many forms of life have taken their rise.

During the long Mesozoic period fish-lizards played the part that whales now play in the economy of the world; and they resembled the latter, not only in general shape, but in the situation of the nostrils (near the eye), and in their teeth and long jaws. But these curious resemblances must not be interpreted to mean that whales and fish-lizards are related to each other. They only show that similar modes of life tend to produce artificial resemblances—just as some whales, in their turn, show a superficial resemblance to fishes.

With regard to the particular form of reptile from which the fish-lizard may have been derived, no certain conclusion has at present been arrived at. This is chiefly from want of fuller knowledge of early forms, such as may have existed in the previous periods known as the Carboniferous and Trias (see Appendix I.). But there are certain features in the skulls, teeth, and vertebræ that suggest a relationship with the Labyrinthodonts, or primæval salamanders that flourished during the above periods, or at least from amphibians more or less closely allied to them. They cannot by any possibility be regarded as modified fishes; for fishes have gills instead of lungs.

The fish-lizards played their part, and played it admirably; but their days were numbered, and the place they occupied has since been taken by a higher type—the mammal. As reptiles, they were eminently a success; but, then, they were only reptiles, and therefore were at last left behind in the struggle for existence, until finally they died out, at the end of the Cretaceous period, when certain important geographical and other changes took place, helping to cause the extinction of many other strange forms of life, as we shall see later on (see p. 147).

They had a wide geographical range; for their remains have been discovered in Arctic regions, in Europe, India, Ceram, North America, the east coast of Africa, Australia, and New Zealand.

In American deposits they are represented by certain toothless forms, to which the name Sauranodon ("toothless lizard") has been given. These have been discovered by Professor Marsh, in the Jurassic strata of the Rocky Mountains. They were eight or nine feet long, and in every other respect resembled Ichthyosaurs. As we have endeavoured to indicate in our illustration, the fish-lizards flourished in seas wherein animal, and doubtless vegetable life was very abundant. Any one who has collected fossils from the Lias of England will have found how full it is of beautiful organic remains, such as corals, mollusca, encrinites, sea-urchins, and other echinoids, fishes, etc.

The climate of this period in Europe was mild and genial, or even semi-tropical. Coral reefs and coral islands varied the landscape. There is just one more point of interest that ought not to be omitted; it refers to the manner in which these reptiles of the Lias age met their deaths, and were thus buried up in their rocky tombs. Sir Charles Lyell and other writers point out that the individuals found in those strata must have met with a sudden death and quick burial; for if their uncovered bodies had been left, even for a few hours, exposed to putrification and the attacks of fishes at the bottom of the sea, we should not now find their remains so completely preserved that often scarcely a single bone has been moved from its right place. What was the exact nature of this operation is at present a matter of doubt.

CHAPTER IV.

THE GREAT SEA-LIZARDS AND THEIR ALLIES.

"The wonders of geology exercise every faculty of the mind—reason, memory, imagination; and though we cannot put our fossils to the question, it is something to be so aroused as to be made to put the questions to one's self."—HUGH MILLER.

THE fish-lizards, described in our last chapter, were not the only predaceous monsters that haunted the seas of the great Mesozoic age, or era. We must now say a few words about certain contemporary creatures that shared with them the spoils of those old seas, so teeming with life. And first among these—as being more fully known—come the long-necked sea-lizards, or Plesiosaurs.

The Plesiosaurus was first discovered in the Lias rocks of Lyme-Regis, in the year 1821. It was christened by the above name, and introduced to the scientific world by the Rev. Mr. Conybeare (afterwards Dean of Llandaff) and Mr. (afterwards Sir Henry) de la Beche. They gave it this name in order to distinguish it from the Ichthyosaurus, and to record the fact that it was more nearly allied to the lizard than the latter.[1] Conybeare, with the assistance of De la Beche, first described it in a now-classic paper read before the Geological Society of London, and published in the *Transactions* of that Society in the year 1821. In a later paper (1824) he gave a restoration

[1] The name is derived from two Greek words—*plesios*, near, or allied to, and *sauros*, a lizard.

of the entire skeleton of Plesiosaurus dolichodeirus; and the accuracy of that restoration is still universally acknowledged. This fine specimen was in the possession of the Duke of Buckingham, who kindly placed it at the disposal of Dr. Buckland, for a time, that it might be properly described and investigated.

A glance at our illustration, Plate III., will show that this strange creature was not inaptly compared at the time to a snake threaded through the body of a turtle.

Dr. Buckland truly observes that the discovery of this genus forms one of the most important additions that geology has made to comparative anatomy. " It is of the Plesiosaurus," says that graphic author, in his *Bridgewater Treatise*, "that Cuvier asserts the structure to have been the most heteroclite, and its characters altogether the most monstrous that have been yet found amid the ruins of a former world. To the head of a lizard it united the teeth of a crocodile ; a neck of enormous length, resembling the body of a serpent ; a trunk and tail having the proportions of an ordinary quadruped ; the ribs of a chameleon, and the paddles of a whale ! Such are the strange combinations of form and structure in the Plesiosaurus—a genus, the remains of which, after interment for thousands of years amidst the wreck of millions of extinct inhabitants of the ancient earth, are at length recalled to light by the researches of the geologist, and submitted to our examination in nearly as perfect a state as the bones of species that are now existing upon the earth."

Perhaps the best way in which we can gain a clear idea of the general characters of a long-necked sea-lizard, as we may call our Plesiosaurus, is by comparing it with the fish-lizard, described in the last chapter. Its long neck and small head are the most conspicuous features. Then we notice the short tail. But if we compare the paddles of these two extinct forms of life, we notice at once certain important differences. In the fish-lizard the bone of the arm was very short, while all the bones of the fore-arm and fingers were modified into little many-sided bodies, and so articulated together as to make the whole limb, or paddle, a solid yet flexible structure. In the long-necked sea-lizard, however, we find a long arm-bone with a club-like shape; and the two bones of the fore-arm are seen to be longer than in the fish-lizard. But a still greater difference shows itself in the bones of the finger, as we look at a fossilised skeleton (or a drawing of one); for the fingers are long and slender, like those of ordinary reptiles.

There are only five fingers, and each finger is quite distinct from the others. This is the reason why the Plesiosaur was considered to depart less from the type of an ordinary reptile, and so received its name. Other remarkable differences present themselves in the shoulders and haunches, but these need not be considered here. The species shown in Fig. 8 had rather a large head. It is obvious that such a long slender neck as these creatures had could not have supported a large head, like that of the fish-lizard. Consequently, we find a striking contrast in the skulls of the two forms. That of the Plesiosaur was short and stout, and therefore such as could easily be supported, as well as rapidly moved about by the long slender neck. Thus we find another simple illustration of the "law of correlation," alluded to on p. 6. The teeth were set in distinct sockets, as they are in crocodiles, to which animals there are also points of resemblance, in the backbone, ribs, and skull. Fig. 7 shows three different types of lower jaws of Plesiosaurs. The one marked C belongs to Plesiosaurus dolichodirus, the species represented in our plate. There were no bony plates in the eye. Professor Owen thinks that they were long-lived. The skin was probably smooth, like that of a porpoise.

The visitor to the geological collection at South Kensington will find a splendid series of the fossilised remains of long-necked sea-lizards. They were mostly obtained from the Lias formation of Street in Somersetshire, Lyme-Regis in Dorset, and Whitby in Yorkshire. Those from the Lias are mostly small, about eight

to ten feet in length. But in the rocks of the Cretaceous period, which was later, are found larger specimens. There is a cast of a very fine specimen from the Upper Lias on the wall of the east corridor (No. 3 on Plan) of the geological galleries at South

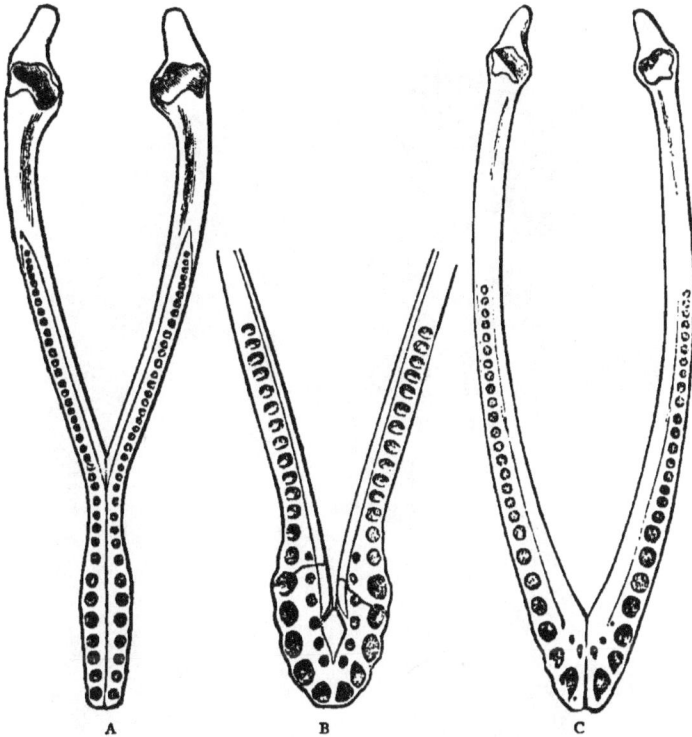

Fig. 7.—Mandibles of Fish-lizards. A, *Peloneustes philarchus* (Seeley); from the Oxford Clay. B, *Thaumatosaurus indicus* (Lydekker); Upper Jurassic of India. C, *Plesiosaurus dolichodirus* (Conybeare); from the Lower Lias, Lyme Regis.

Kensington, which is twenty-two feet long. But some of the Cretaceous forms, both in Europe and America, attained a length of forty feet, and had vertebræ six inches in diameter. The bodies of the vertebræ, or "cup-bones," are either flat or slightly concave, showing that the backbone as a whole was less flexible than in the fish-lizards.

It may be mentioned here that Mr. Smith Woodward, of the Natural History Museum, recently showed the writer a fossil Plesiosaur that is being set up in the formatore's shop, in the same manner that a recent skeleton might be. In this, and many

FIG. 8.—*Plesiosaurus macrocephalus.*

other ways, the guardians of the national treasure-house are endeavouring to make the collection intelligible and interesting to the general public. Specimens of extinct animals thus set up, give one a much better idea than when the bones are all lying huddled together on a slab of rock. But it is not always possible to get the bones entirely out of their rocky bed, or matrix.

Great credit is due to Mr. Alfred N. Leeds, of Eyebury, who has disinterred the separate bones of many distinct skeletons of Plesiosaurs from Oxford Clay strata near Peterborough.

It will be remembered that the long and powerful tail of the fish-lizard was its principal organ of propulsion through the water; and that, consequently, the paddles only played a secondary part. They were small, but amply large enough for the work they had to perform. But our long-necked sea-lizards possessed very short tails. What, then, was the consequence? Obviously that the paddles had all the more work to do. They were the chief swimming organs. The vertebræ of this short tail show that it probably was highly flexible, and could move rapidly from side to side; but, for all that, its use as a propeller would not be of much importance. We see now why the paddles are so long and powerful, like two pairs of great oars, one pair on each side of the body. In a fossil skeleton you will notice the flattened shape of the arm-bone (or humerus), and of the thigh-bone (or femur). This gave breadth to the paddles, and made them more efficient as swimming organs. They give no indication of having carried even such imperfect claws as those of turtles and seals, and therefore we may conclude that the Plesio-saur was far more at home in the water than on land, and it seems probable that progression on land was impossible.

The tail was probably useful as a rudder, to steer the animal when swimming on the surface, and to elevate or depress it in ascending and descending through the water. Like the fish-lizard, this creature was an air-breather, and therefore was obliged occasionally to visit the surface for fresh supplies of air. But probably it possessed the power of compressing air within its lungs, so that the frequency of its visits to the surface would not be very great.

From the long neck and head, situated so far away from the paddles, as well as for other reasons, it may be concluded that this creature was a rapid swimmer, as was the Ichthyosaurus.

Although of considerable size, it probably had to seek its food, as well as its safety, chiefly by artifice and concealment. The fish-lizard, its contemporary, must have been a formidable rival and a dangerous enemy, whom to attack would be unadvisable.

Speaking of the habits of the long-necked sea-lizard, Mr. Cony-beare, in his second paper, already alluded to, says, " That it was aquatic, is evident from the form of its paddles ; that it was marine, is almost equally so, from the remains with which it is uni-versally associated ; that it may occasionally have visited the shore, the resemblance of its extremities to those of the turtle may lead us to conjecture ; its motion, however, must have been very awkward on land ; its long neck must have impeded its progress through the water, presenting a striking contrast to the organisa-tion which so admirably fits the Ichthyosaurus to cut through the waves.

" May it not therefore be concluded (since, in addition to these circumstances, its respiration must have required frequent access of air) that it swam upon or near the surface, arching back its long neck like the swan, occasionally darting it down at the fish which happened to float within its reach ?   It may, perhaps, have lurked in shoal-water along the coast, concealed among the sea-weed, and, raising its nostrils to a level with the surface from a considerable depth, may have found a secure retreat from the assaults of dangerous enemies ; while the length and flexibility of its neck may have compensated for the want of strength in its jaws and its incapacity for swift motion through the water, by the suddenness and agility of the attack which they enabled it to make on every animal fitted for its prey, which came within its extensive sweep."

More than twenty species of long-necked sea-lizards are known to geologists.

Professor Owen, in his great work on *British Fossil Reptiles,* when describing the huge Plesiosaurus dolichodirus from Dorset, suggests that the carcase of this monster, after it sank to the bottom of the sea, was preyed upon by some carnivorous animal (perhaps sharks). It seems, he says, as if a bite of the neck had pulled out of place the eighth to the twelfth vertebræ. Those at the base of the neck are scattered and dispersed as if through more "tugging and riving." So with regard to its body, probably some hungry creature had a grip of the spine near the middle of the back, and pulled all the succeeding vertebræ in the region of the hind limbs. Thus we get a little glimpse of scenes of violence that took place at the bottom of the bright sunny seas of the period when the clays and limestones of the Lias rocks were being deposited in the region of Lyme-Regis.

As time went on, these curious reptiles increased in size, until, in the period when our English chalk was being formed (Cretaceous period), they reached their highest point (see p. 147). After that they became extinct—whether slowly or somewhat suddenly we cannot tell.

Until more is known of the ancient life of the earth, it will not be possible to say with certainty what were the nearest relations of the long-necked sea-lizards. They first appear in the strata of the New Red Sandstone, which is below the Lias. Certain little reptiles, about three feet long, from the former rocks, known as Neusticosaurus and Lariosaurus, seem to be rather closely related to the creatures we are now considering, and to connect them with another group, namely, the Pliosaurs. They were partly terrestrial and partly aquatic; but it is not easy to say whether their limbs had been converted into true paddles or not. At any rate, there is every reason to believe that the long-necked sea-lizards were descended from an earlier form of land reptile. They gradually underwent considerable modifications, in order to adapt themselves to an aquatic life. We noticed that the same conclusion has been arrived at with regard to the fish-lizards. Both these extinct groups, therefore, present an interesting analogy to whales, which are now considered to have been derived, by a like series of changes, from mammals that once walked the earth.

The Plesiosaur presents, on the one hand, points of resemblance to turtles and lizards,—on the other hand, to crocodiles, whales, and, according to some authorities, even the strange Ornithorhynchus. But it will be very long before its ancestry can be made known. In the mean time, we must put it in a place somewhere near the fish-lizards, and leave posterity to complete what has at present only been begun. It must, however, be borne in mind that some of the above resemblances are purely accidental, and not such as point to relationship. Because their flippers are like those of a whale, it does not mean that Plesiosaurs are related to modern whales. It only means that similar habits tend to produce accidental resemblances — just as the whales and porpoises, in their turn, resemble fishes. To make torpedoes go rapidly through the water, inventors have given them a fish-like shape ;—in the same way the early forms of mammals, from which whales are descended, gradually adapted themselves to a life in the water, and so became modified to some extent to the shapes of fishes.

The Pliosaurs, above mentioned, are evidently relations, but with short necks instead of long ones. They had enormous heads and thick necks. Fine specimens of their huge jaws, paddle-bones, etc., may be seen at the end of the reptile-gallery at Cromwell Road. One of the skulls exhibited there is nearly six feet long, while a hind paddle measures upwards of six and a half feet in length, of which thirty-seven inches is taken up by the thigh-bone alone. The teeth at the end of the jaws are truly enormous. One tooth, from a deposit known as the Kimmeridge Clay, is nearly a foot long from the tip of the crown to the base of the root. In some, the two jaw-bones of the lower jaw are partly united, as in the sperm-whale or cachalot. Creatures so armed must have been very destructive.

CHAPTER V.

" What we know is but little ; what we do not know is immense."—LA
PLACE.

WAS there ever an age of dragons ? Tradition says there was ;
but there is every reason to believe that the fierce and blood-
thirsty creatures, of which such a variety present themselves, are
but creations of the imagination,—useful in their way, no doubt,
as pointing a moral or adorning a tale, but, nevertheless,
wholly without foundation in fact. The dragon figures in the
earliest traditions of the human race, and crops up again in full
force in European mediæval or even late romance.

In ancient Egyptian mythology, Horus, the son of Isis, slays
the evil dragon. In Greece, the infant Hercules, while yet in his
cradle, strangles deadly snakes; and Perseus, after engaging in
fierce struggle with the sea-monster, slays it, and rescues Andro-
meda from a cruel death. In England, we have the heroic legend
of St. George and the Dragon depicted on our sovereigns. But it
is easy to see a common purpose running through these legends.
They are considered by many to be solar myths, and have a moral
purpose. The dragons or snakes are emblems of darkness and
evil ; the heroes emblems of light, and so of good. The triumph
of good over evil is the theme they were intended to illustrate.
The dragons, then, are clearly products of the imagination, based,
no doubt, on the huge and uncouth reptiles of the present human
era, such as crocodiles, pythons, and such creatures.

Amidst much diversity there is yet a strange similarity in the dragons that figure in the folk-lore of Eastern and Western peoples.   Probably our European traditions were brought by the tribes which, wave after wave, poured in from Central Asia.

They are, for the most part, unnatural beasts, breathing out fire, and often endowed with wings, while at the same time possessing limbs ending in cruel claws, fitted for clutching their unfortunate victims.   The wings seem, to say the least, very much in the way.   Poisonous fangs, claws, scaly armour, and a long pointed tail were all very well,—but wings are hardly wanted, unless to add one more element of mystery or terror.   Some, however, are devoid of wings : the Imperial Japanese dragons showing no sign of such appendages.   The Temple Bar griffin is a grim example of a winged monster.   Nevertheless, in spite of all the manifest absurdities of the dragons of various nations and times, geology reveals to us that there once lived upon this earth reptiles so great and uncouth that we can think of no other but the time-honoured word " dragon " to convey briefly the slightest idea of their monstrous forms and characters.

So there is some truth in dragons, after all.   But then we must make this important reservation—viz. that the days of these dragons were long before the human period; they flourished in one of those dim geological ages of which the rocks around us bear ample records.

It is a strange fact that human fancy should have, in some cases at least, created monsters not very unlike some of those antediluvian animals that have, during the present century, been discovered in various parts of Europe and America.   Some unreasonable persons will have it that certain monstrous reptiles of the Mesozoic era, about to be described, must have somehow managed to survive into the human period, and so have suggested to early races of men the dragons to which we have alluded. But there is no need for this untenable supposition.   By a free blending together of ideas culled from living types of animals it would be very easy to construct no small variety of dragons; and so we may believe this is what the ancients did.

Having said so much of dragons in general, let us proceed to consider those both possible and real monsters revealed of late years by the researches of geologists. For this purpose we shall devote the present and two following chapters to the consideration of a great and wonderful group of fossil reptiles known as Dinosaurs. The strange fish-lizards and sea-lizards previously described were the geological contemporaries of a host of reptiles, now mostly extinct, which inhabited both the lands and waters of those periods known as the Triassic, Jurassic, and Cretaceous, which taken together represent the great Mesozoic, formerly called the Secondary, era.

The announcement by Baron Cuvier—the illustrious founder of Palæontology—that there was a period when our planet was inhabited by reptiles of appalling magnitude, with many of the features of modern quadrupeds, was of so novel and startling a character as to require the prestige of even his name to obtain for it any degree of credence. But subsequent discoveries have fully confirmed the truth of his belief, and the " age of reptiles " is no longer considered fabulous. This expression was first used by Dr. Mantell as the title of a paper published in the *Edinburgh Philosophical Journal* in 1831, and serves to remind us that reptilian forms of life were once the ruling class among animals.

The Dinosaurs are an extinct order comprising the largest terrestrial and semi-aquatic reptiles that ever lived; and while some of them in a general way resembled crocodiles, others show in the bony structures they have left behind a very remarkable and interesting resemblance to birds of the ostrich tribe. This resemblance shows itself in the pelvis, or bony arch with which the hind limbs are connected in vertebrate or back-boned animals, and in the limbs themselves. This curious fact, first brought into notice by Professor Huxley, has been variously interpreted by anatomists; some concluding, with Professor Huxley, that birds are descended from Dinosaurs; while others, with Professor Owen, consider the resemblance accidental, and in no way implying relationship. Huxley has proposed the name Ornitho-scelida, or bird-legged, for these remarkable reptiles.

Dinosaurs must have formerly inhabited a large part of the primæval world; for their remains are found, not only in Europe, but in Africa, India, America, and even in Australia; and the geologist finds that they reigned supreme on the earth throughout the whole of the great Mesozoic era.

Their bodies were, in some cases, defended by a formidable coat of armour, consisting of bony plates and spines, as illustrated by the case of Scelidosaurus (p. 105), thus giving them a decidedly dragon-like appearance. The vertebræ, or bony segments of the back-bone, generally have their centra hollow on both sides, as in the Ichthyosaurus; but in the neck and tail they are not unfrequently hollow on one side and convex on the other. In some of the largest forms the vertebræ are excavated into hollow chambers. This is apparently for the sake of lightness; for a very large animal with heavy solid bones would find it difficult to move freely. In this way strength was combined with lightness.

All the Dinosaurs had four limbs, and in many cases the hind pair were very large compared to the fore limbs. They varied enormously in size, as well as in appearance. Thus certain of the smaller families were only two feet long and lightly built; while others were truly colossal in size, far out-rivalling our modern rhinoceroses and elephants.

The limbs of Cetiosaurus, for example, or of Stegosaurus, remind us strikingly of those of elephants. The celebrated Von Meyer was so struck with this likeness that he proposed the name Pachypoda for them, which means thick-footed. Professor Owen opposed this name; for it was misleading, and only applied to a few of them. He therefore proposed the name we have already been using, viz. Dinosauria,[1] and this name has been generally retained. We are thus led to connect them with lizards and crocodiles, rather than with birds or quadrupeds. The strange and curiously mixed characters of the old-fashioned reptiles is forcibly illustrated by these differences of opinion among leading naturalists. Professor Seeley, another living authority, refuses to consider them as reptiles, at least in the ordinary sense of the word.

Extinct forms of life are often so very different to the creatures inhabiting the world of to-day, that naturalists find it a hard task to assign them their places in the animal kingdom. The classes, orders, and families under which living forms are grouped are often found inadequate for the purpose, so much so that new orders and new families require to be made for them ; and then it is often quite impossible to determine the relations of these new groups to the old ones we are accustomed to. Dinosaurs offer a good example of this difficulty. Were they related to ancient crocodiles ? No one can say for certain ; but it is quite possible, and even probable. Again, did certain long-legged Dinosaurs eventually give rise by evolution to the running birds, ostriches, emeus, etc. ? This, although supported by weighty authority, is a matter of speculation : we ought to be very careful in accepting such conclusions. It may perhaps be safer to look upon the ancestry of birds as one of those problems on which the oracle of science cannot at present declare itself.

Various attempts have been made to classify Dinosaurs, and arrange them in family groups ; but, considering our imperfect knowledge, it will be wise to regard all such attempts as purely temporary and provisional, although in some ways convenient. Professor Marsh, of Yale College, U.S., whose wonderful discoveries in the far West have attracted universal attention, has grouped the Dinosaurs into five sub-orders. It will, however, be sufficient for our purpose if we follow certain English authorities who

---

[1] Greek—*deinos*, terrible ; *sauros*, lizard.

F

divide them into three groups—taking the names given by Professor Marsh, only running together some which he would separate.

We shall first consider the very interesting and huge forms included in his sub-order the Sauropoda, or lizard-footed Dinosaurs. Various parts of the skeletons, such as vertebræ, leg-bones, etc., of these cumbrous beasts have long been known in this country; but Professor Marsh was the first person to discover a complete skeleton.

We shall, therefore, now turn our attention to the bony framework of the huge Brontosaurus (Fig. 9), a vegetable-feeding lizard. But it will be necessary to completely lay aside all our previous notions taken from lizards of the present day, with their short legs and snake-like scaly bodies, before we can come to any fair conclusion with regard to this monstrous beast.

It was nearly sixty feet long, and probably when alive weighed more than twenty tons! that it was a stupid, slow-moving reptile, may be inferred from its very small brain and slender spinal cord. By taking casts of the brain-cavities in the skulls of extinct animals, anatomists can obtain a very good idea of the nature and capacity of their brains; and in this way important evidence is obtained, and such as helps to throw light upon their habits and general intelligence. No bony plates or spines have been discovered with the remains of this monster; so that we are driven to conclude that it was wholly without armour: and, moreover, there seem to be no signs of offensive weapons of any kind.

Professor Marsh concludes that it was more or less amphibious in its habits, and that it fed upon aquatic plants and other succulent vegetation. Its remains, he says, are generally found in localities where the animal had evidently become mired, just as cattle at the present day sometimes become hopelessly fixed in a swampy place on the margin of a lake or river (see p. 19). Each track made by the creature in walking occupied one square yard in extent!

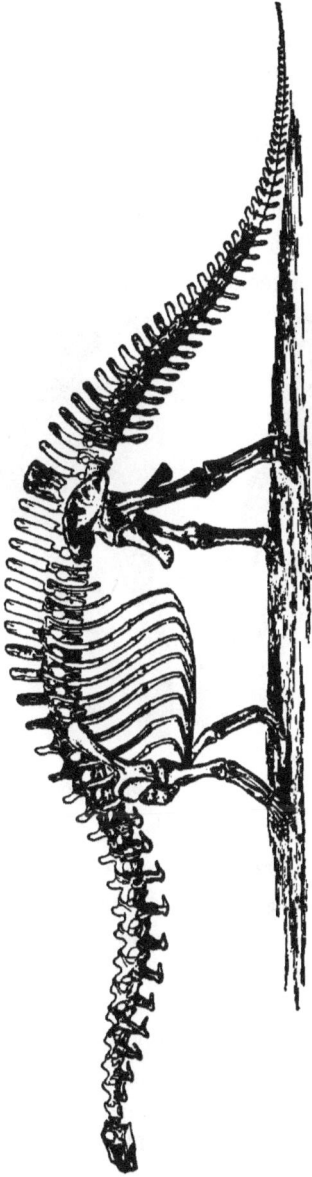

FIG. 9.—Restored skeleton of *Brontosaurus excelsus.* (After Marsh.)

The remarkably small head is one of the most striking features of this Dinosaur, and presents a curious contrast to the large and formidable skulls possessed by some other forms to be described further on.　But it is clear that no animal with such a long neck as this creature had could have borne the weight of a heavy skull. Short thick necks and heavy skulls always go together.　Indeed, the weight of the long neck itself would have been serious had it not been for the fact that the vertebræ in this part of the skeleton, and as far as the region of the tail, have large cavities in the sides of

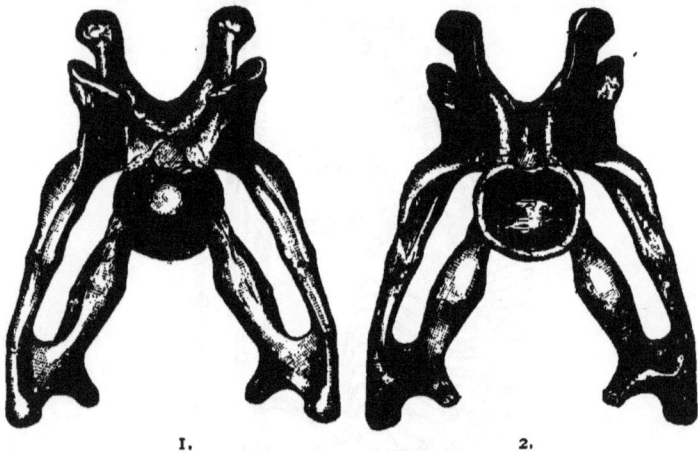

1.　　　　　　　　　　　　　　2.

FIG. 10.—Neck vertebræ of *Brontosaurus.*
1. Front view.　2. Back view.

the centra.　This cavernous structure of the vertebræ gradually decreases towards the tail.　The cavities communicated with a series of internal cavities which give a kind of honeycombed structure to the whole vertebra.　This arrangement affords a combination of strength and lightness in the massive supports required for the huge ribs, limbs, and muscles, such as could not have been provided by any other plan.　(See Fig. 10.)

The body of the Brontosaur was comparatively short, with a fairly large paunch (see restoration, Plate IV.).　The legs and feet were

A GIGANTIC DINOSAUR, BRONTOSAURUS EXCELSUS.
Length nearly 60 feet.

strong and massive, and the limb-bones solid. As if partly in order to balance the neck, we find a long and powerful tail, in which the vertebræ are nearly all solid. In most Dinosaurs the fore limbs are small compared to the hind limbs—*e.g.* Megalosaurus, Iguanodon, and Scelidosaurus,—but here we find them unusually large. In this case, then, it is hardly possible that the creature walked upon its hind legs, as many of the Dinosaurs did. But, at the same time, we may believe that occasionally it assumed a more erect position; and may not the light hollowed structure of the vertebræ in the fore part of the body, already alluded to, have imparted such lightness as made it possible for the creature to assume such attitudes ? There can be little doubt but that many other fierce and formidable Dinosaurs were living at the same time and in the same region with Brontosaurus, whose remains are found in the Jurassic rocks of Colorado (Atlantosaurus beds).

How this apparently helpless and awkward animal escaped in the struggle for existence it is not easy to conjecture; but since there is reason to believe it was more or less at home in the water, and could use its powerful tail in swimming, we may perhaps find a way out of the difficulty by supposing that, when alarmed by dangerous flesh-eating foes, it took to the water, and found discretion to be the better part of valour. Although apparently stupid, the Brontosaur probably possessed a good deal of cunning, and we can fancy it stretching its long neck above reeds, ferns, and cycads to get a view of the approaching enemy.

The Sauropoda, or lizard-footed Dinosaurs, show in many ways a decided approach to a simple or generalised crocodile ; so much so, that Professor Cope is inclined to include crocodiles and sauropodous Dinosaurs in the same order. Still, there are important differences in other members of this sub-order. Unfortunately, our knowledge is at present rather limited, owing to the want of complete skeletons. Vertebræ, limb-bones, skulls, and teeth have all been discovered through the zeal and energy of

Professor Marsh and his comrades, in the far west of America, as well as by the researches of English geologists, assisted by the labours of many ardent collectors of fossils, in this country. Some of these may now be briefly considered.

In Plate V. we have endeavoured to give some idea of a huge thigh-bone (femur) belonging to the truly gigantic Dinosaur called Atlantosaurus. It is six feet two inches long, and a cast of it may be seen in the fossil reptile gallery of the British Museum of Natural History (Wall-case No. 3). It should be mentioned, however, that the original specimen is partly restored, so that its exact length to an inch or so is not quite certain. In our illustration it is shown to be a little taller, when placed upright, than a full-grown man. Professor Marsh, the fortunate discoverer of this wonderful bone, calculates that the Atlantosaurus must have attained a length of over eighty feet! and, assuming that it walked upon its hind feet, a height of thirty feet!

It doubtless fed upon the luxuriant foliage of the sub-tropical forests, portions of which are preserved with its remains. Besides this thigh-bone, Professor Marsh has procured specimens of vertebræ from the different parts of the vertebral column ; but no skull or teeth. The vertebræ are hollowed out much in the same way as those of Brontosaurus. The fore limbs were large, as in the latter animal ; and the extremities of the limbs were provided with claws. Taking all present evidence, it appears that the Atlantosaurus bore a general resemblance to its smaller contemporary. We can therefore form a fairly good idea of its aspect and proportions.

The same Jurassic strata from the Rocky Mountains have yielded remains of another big Dinosaur, belonging to the same family. This genus, which has been named the Apatosaurus, is represented by a nearly complete skeleton, in the Yale College Museum ; and is fortunately in an excellent state of preservation. Another species, of smaller size, though not so complete, adorns the same collection. This was about thirty feet long, and is known as Apatosaurus grandis.

THIGH-BONE OF THE LARGEST OF THE DINOSAURS, ATLANTOSAURUS.
From a cast in the Natural History Museum. Length 6 feet 2 inches.
PLATE V.

Morosaurus, another important genus, is known from a large number of individuals discovered in the now famous Atlantosaurus beds of Colorado, including one nearly complete skeleton. The head of this creature was small; the neck elongated; and the vertebræ of the neck are lightened by deep cavities in their centra, similar to those in birds of flight. The tail, also, was long. When alive, this Dinosaur was about forty feet in length. It probably walked on all fours; and in many other respects was very unlike a typical Dinosaur. The brain was small, and it must have been sluggish in all its movements. The nearly complete remains of Morosaurus grandis were found together in a very good state of preservation in Wyoming, and many of the bones lay just in their natural positions.

Diplodocus, of which several incomplete specimens have been discovered, was intermediate in size between Atlantosaurus and Morosaurus, and may have reached when living, a length of forty or fifty feet. Its skull was of moderate size, with slender jaws. The teeth were weaker than those of any other known Dinosaur, and entirely confined to the front of the jaws. Professor Marsh concludes from the teeth that Diplodocus was herbivorous, feeding on succulent vegetation, and that it probably led an aquatic life. Fig. 11 shows its skull.

The remains of this interesting Dinosaur (Brontosaurus), which in several ways differs from other members of the " lizard-footed " group, were found in Upper Jurassic beds, near Cãnon City, Colorado. A second smaller species was also discovered near Morrison, Colorado. All the remains lay in the Atlantosaurus beds. These strata—the tomb in which Nature has buried up so many of her dragons of old time—can be traced for several hundred miles on the flanks of the Rocky Mountains, and are always to be known by the bones they contain. They lie above the Triassic strata and just below the Sandstone of the Dakota group. Some have regarded them as of Cretaceous age; but, judging from their fossils, there can be but little doubt that they were deposited during the Jurassic period—probably in an old estuary. They consist of shale and sandstone.

Besides the numerous Dinosaurs, Professor Marsh's colleagues have found abundant remains of crocodiles, tortoises, and fishes, with one Pterodactyl, a flying reptile (see chap. viii.), and several small marsupials. The wonderful collection of American Jurassic Dinosaurs in the Museum of Yale College includes the remains of several hundred individuals, many of them in excellent pre-

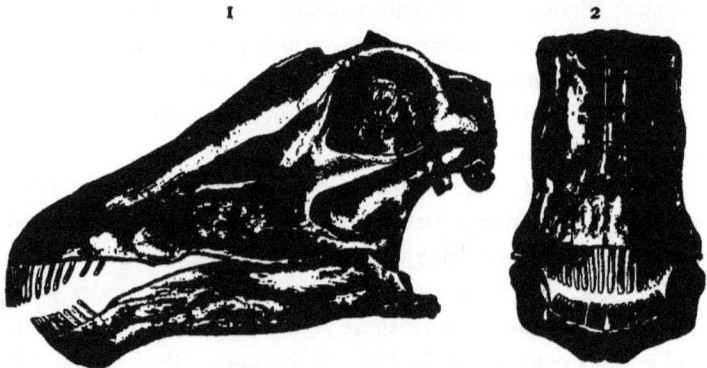

FIG. 11.—Head of *Diplodocus*.
1. Side view. 2. Front view.

servation, and has afforded to Professor Marsh the material for his classification already alluded to.

ENGLISH DINOSAURS OF THE LIZARD-FOOTED GROUP.

Unfortunately, there are at present no complete skeletons known of English Dinosaurs related to the American forms above de-scribed. But, since the English fossils were first in evidence by many years, and Marsh's discoveries have confirmed in a remark-able way conclusions drawn by Owen, Huxley, Hulke, and Seeley, and others from materials that were rather fragmentary, it may be worth while to give some account of these remains and the inter-pretations they have received.

Dr. Buckland, in his *Bridgewater Treatise*, 1836, referred to a limb-bone in the Oxford Museum, from the great Oölite formation near Woodstock, which was examined by Cuvier, and pronounced to have once belonged to a whale; also a very large rib, which seemed whale-like. In 1838 Professor Owen, when collecting materials for his famous *Report on the Fossil Reptiles of Great Britain*, inspected this remarkable limb-bone, and could not match it with any bones known among the whale tribe; and yet its structure, where exposed, was like that of the long bone (humerus) of the paddle of a whale. Later on, he abandoned the idea that it once belonged to a whale, and it was thought that the extinct animal in question might have been a reptile of the crocodilean order. In time, a fine series of limb-bones and vertebræ was added to the Oxford Museum by Professor Phillips (Dr. Buckland's successor at Oxford), who pronounced them to be Dinosaurian. The name "Cetiosaurus"[1] (or Whale-lizard), originally given by Owen, was unfortunate, because there is really nothing whale-like about it, except a certain coarse texture of some of the bones.

In 1848 Dr. Buckland announced the discovery of another limb-bone (a femur), which Owen referred to Cetiosaurus; it was four feet three inches in length. Between 1868 and 1870, however, a considerable portion of a skeleton was discovered in the same formation at Kirtlington Station, near Oxford. These remains were the subject of careful examination by Professors Owen and Phillips. The femur this time was five feet four inches long. Their studies threw much light on the nature and habits of Cetiosaurus.

Although showing in many ways an approach to the crocodile type of reptile, yet it was perceived from the nature of the limbs that they were better fitted for walking on land than are those of a crocodile, with its sprawling limbs. Still, Professor Owen was careful to point out that the vertebræ of its long tail indicate suitability as a powerful swimming organ, and concluded that the

[1] Greek—*ketion*, whale; *sauros*, lizard.

creature was more aquatic than terrestrial in its habits. Plaster casts of the limb-bones may be seen at the British Museum of Natural History, side by side with the huge Atlantosaurus cast sent by Professor Marsh.

The Kimmeridge clay of Weymouth has yielded a huge arm-bone (or humerus), nearly five feet long ; and from Wealden strata of Sussex and the Isle of Wight vertebræ have been collected. Altogether we have remains of Cetiosaurus from at least half a dozen counties. Unfortunately, no specimen of a skull has yet been found, and only two or three small and incomplete teeth, which may possibly have belonged to some other animal. Professor Owen estimated the length of the trunk and tail of the creature to have been thirty-five or thirty-six feet; but in the absence of further evidence it was not possible to form any conclusion as to its total length. It is evident that Cetiosaurus was closely allied to the American Brontosaurus (p. 69) ; and so these earlier English discoveries have gained much in interest from the light thrown upon them by Professor Marsh's huge Saurian.

Another English Saurian of this group was the Ornithopsis, from Wealden strata in the Isle of Wight, which has been the subject of careful study by Mr. Hulke and Professor Seeley. Their conclusions, based on the examination of separate portions of the skeleton (such as vertebræ), have been singularly confirmed by the discovery of Brontosaurus.

In Ornithopsis the vertebræ of the neck and back, though of great size, were remarkably light, and yet of great strength. One of the vertebræ of the back had a body, or centrum, ten inches long. Hoplosaurus and Pelosaurus were evidently reptiles closely allied to the above types ; but at present are so imperfectly known that we need not consider them here.

CHAPTER VI.

DINOSAURS (*continued*).

"Fossils have been eloquently and appropriately termed 'Medals of Creation.'"—DR. MANTELL.

WHEN any tribe of plants or animals becomes very flourishing, and spreads over the face of the earth, occupying regions far apart from one another, where the geographical and other conditions, such as climate, are unlike, its members will inevitably develop considerable differences among themselves.

During the great Mesozoic period, Dinosaurs spread over a large part of the world ; they became very numerous and powerful. Just as the birds and beasts (quadrupeds) of to-day show an almost endless variety, according to the circumstances in which they are placed, so that great and powerful order of reptiles we are now considering ran riot, and gave rise to a variety of forms, or types. Those described in the last chapter were heavy, slow-moving Dinosaurs, of great proportions, and were all herbivorous creatures, apparently without weapons of offence or defence.

The group Theropoda, or "beast-footed" Dinosaurs, that partly form the subject of the present chapter, were all flesh-eating animals ; and, as we shall discover from their fossilised remains, were of less size, and led active lives. In fact, they acted in their day the part played by lions and tigers to-day.

In the year 1824 that keen observer and original thinker, the Rev. Dr. Buckland, described to the Geological Society of London some remains of a very strange and formidable reptile found in the Limestone of Stonesfield, near Woodstock (about twelve miles from Oxford). This rock, known as "Stonesfield slate" from its property of splitting up into thin layers, has long been celebrated for its fossil remains, and from it have also been obtained the bones of some early mammals. It is a member of the Lower Oölitic group.

The portions of skeleton originally discovered consisted of part of a lower jaw, with teeth, a thigh bone (femur), a series of vertebræ of the trunk, a few ribs, and some other fragments. The name Megalosaurus,[1] or "great lizard," suggested itself both to Dr. Buckland and Baron Cuvier, because it was evident from the size of the bones that the creature must have been very big. It is true these bones were not found together in one spot; but Professor Owen came to the conclusion that they all belonged to the same species.

No entire skeleton of the Megalosaur has ever been found, but there was enough material to enable Dr. Buckland, Professor Owen, and Professor Phillips to form a very fair idea of its general structure. It should be mentioned here that Dr. Mantell, the enthusiastic geologist to whose labours palæontologists are greatly indebted, had previously discovered similar teeth and bones in the Wealden strata of Tilgate Forest. Sherborne, in Dorset, is another locality which has yielded a fine specimen of parts of both jaws with teeth. A cast of this may be seen in the geological collection at South Kensington. It was found in the Inferior Oolite (Wall-case IV.); the original specimen lies in the museum of Sherborne College. Remains of Megalosaurus have also been found at the following places: Lyme-Regis and Watchet (in the Lias); near Bridport (in Inferior Oolite); Enslow Bridge (upper part of the Great Oolite and Forest Marble Beds); Weymouth (in Oxford Clay); Cowley and Dry Sandford (in the Coral Rag); Malton in Yorkshire (in Coralline Oolite); also in Normandy. They have also been found in Wealden strata.

---

[1] Greek—*megas*, great; *sauros*, lizard.

The portion of a lower jaw in the Oxford Museum is twelve inches long, with a row of nine teeth, or sockets for teeth. The structure of the teeth leaves no doubt as to the carnivorous habits of the creature. With a length of perhaps thirty feet, capable of free and rapid movement on land, with strong hind limbs, short head, with long pointed teeth, and formidable claws to its feet, the Megalosaur must have been without a rival among the carnivorous reptiles on this side of the world. It probably walked for the most part on its hind legs, as depicted in our illustration, and Professors Huxley and Owen, on examining the bones in the Oxford Museum, were much impressed with the bird-like character of some parts of the skeleton, showing an approach to the ostrich type. The form of the teeth, as pointed out by Dr.

FIG. 12.—Lower jaw-bone of Megalosaurus, with teeth.

Buckland, exhibits a remarkable combination of contrivances. When young and first protruding above the gum, the apex of the tooth presented a double cutting edge of serrated enamel ; but as it advanced in growth its direction was turned backwards in the form of a pruning knife, and the enamelled sawing edge was continued downwards to the base of the inner and cutting side, but became thicker on the other side, obtaining additional strength when it was no longer needed as a cutting instrument (Fig. 12).

The genus Megalosaurus—now rendered classic through the labours of Professors Buckland, Phillips, and Owen—may be regarded as the type of the carnivorous Dinosaurs ; and it affords an excellent and instructive instance of the gradual restoration of the skeleton of a new monster from more or less fragmentary remains. Certain very excusable errors were at first made in the restoration, but these have since been rectified by a comparison with the allied American forms, such as Allosaurus, of which nearly entire skeletons have of late been discovered in strata of Jurassic age—in fact, the same rock in Colorado as that in which the huge Atlantosaurus bones lay hid. The accompanying woodcut (Fig. 13) shows how the skeleton has been restored in the light of these later discoveries of Professor Marsh. The large bones of the limbs of these formidable flesh-eating monsters were hollow, and many of the vertebræ, as well as some of those of the feet,

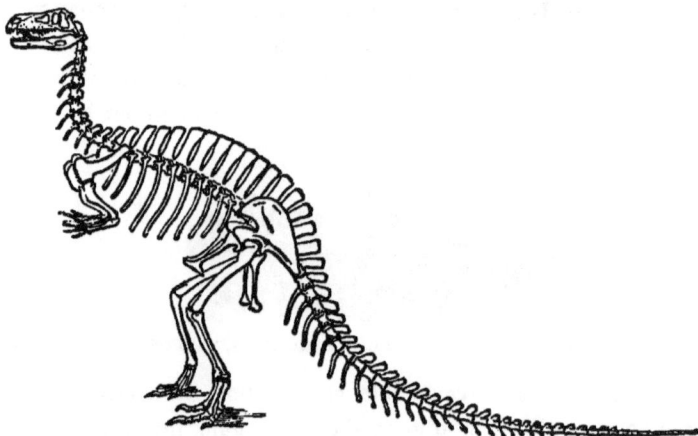

FIG. 13.—Skeleton of Megalosaurus, restored. (After Meyer.)

contained cavities, or were otherwise lightened in order to give the creature a greater power of rapid movement.

It is not very difficult to imagine a Megalosaur lying in wait for his prey (perhaps a slender, harmless little mammal of the ant-eater type) with his hind limbs bent under his body, so as to bring the heels to the ground, and then with one terrific bound from those long legs springing on to the prey, and holding the mammal tight in its clawed fore limbs, as a cat might hold a mouse. Then the sabre-like teeth would be brought into action by the powerful jaws, and soon the flesh and bones of the victim would be gone ! (See Plate VI.)

A CARNIVOROUS DINOSAUR, MEGALOSAURUS BUCKLANDI.

Length about 25 feet.

As we remarked before, the carnivorous Dinosaurs were the lions and tigers of the Mesozoic era, and, what with small mammals and numerous reptiles of those days, it would seem that they were not limited in their choice of diet.

It is a question not yet decided whether Dinosaurs laid eggs as most modern reptiles do, or were viviparous like quadrupeds ; but Professor Marsh thinks there are reasons for the latter supposition.

During the early part of the Mesozoic era, at the period known as the Triassic (New Red Sandstone), Dinosaurs flourished vigorously in America, developing a great variety of forms and sizes. Although but few of their bones have as yet been discovered in those rocks, they have left behind unmistakable evidence of their presence in the well-known footprints and other impressions upon the shores of the waters which they frequented.[1] The Triassic Sandstone of the Connecticut Valley has long been famous for its fossil footprints, especially the so-called " bird-tracks," which are generally supposed to have been made by birds, the tracks of which they certainly appear to resemble. But a careful investigation of nearly all the specimens yet discovered has convinced Professor Marsh that these fossil impressions were not made by birds (see Fig. 14). Most of the three-toed tracks, he thinks, were made by Dinosaurs, who usually walked upon their hind feet alone, and only occasionally put to the ground their small fore limbs. He has detected impressions of the latter in connection with nearly all the larger tracks of the hind limbs. These double impressions are just such as Dinosaurs would make ; and, since the only characteristic bones yet found in the same rocks belong to this order of reptiles, it is but fair to attribute all these footprints to Dinosaurs, even where no impressions of fore feet have been detected, *until* some evidence of birds is forthcoming. The size of some of these impressions, as

[1] Since the above was written, Professor Marsh has described, in *The American Journal of Science* for June, 1892, several more or less complete skeletons of Triassic Dinosaurs, lately found, and now in the Yale College Museum. This is an important discovery.

well as the length of stride they indicate, is against the idea of their having been made by birds. Some of them, for instance, are twenty inches in length, and four or five feet apart! The foot of the African ostrich is but ten inches long, so we must fall back on

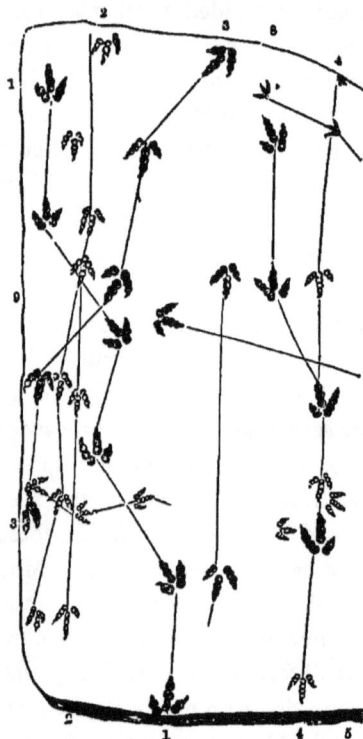

FIG. 14.—Portion of a slab of New Red Sandstone, from Turner's Falls, Massachusetts, U.S., covered with numerous tracks, probably of Dinosaurs This specimen is now in the Natural History Museum. The separate tracks are indicated by the numbers. (After Hitchcock.)

the Dinosaurs for an explanation. However, it is quite possible that some of the smaller impressions were made by birds.

There is at South Kensington a fine series of these and other specimens of fossil footprints (Gallery No. XI., Wall-cases 8–10). The surface of one large slab in the geological collection is eight feet by six feet, and bears upwards of seventy distinct impressions disposed in several tracks, as shown in Fig. 14. The lines were added by Dr. Hitchcock, who has published full descriptions in order to show the direction and disposition of the tracks.

In a presidential address to the Geological Society, Sir Charles Lyell, speaking of the Connecticut Sandstone and its impressions, said, "When I first examined these strata of slate and sandstone near Jersey City, in company with Mr. Redfield, I saw at once from the ripple-marked surface of the slabs, from the casts of

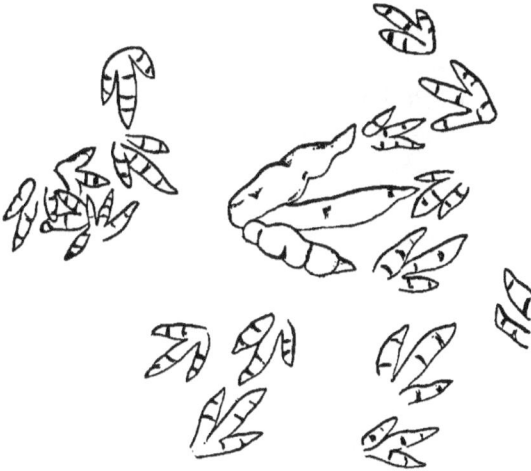

FIG. 15.—Portion of a slab, with tracks. (After Hitchcock.)

cracks, the marks of rain-drops, and the embedded fragments of drift-wood, that these beds had been formed precisely under circumstances most favourable for the reception of impressions of the feet of animals walking between high and low water. In the prolongation of the same beds in the Valley of Connecticut, there have been found, according to Professor Hitchcock, the footprints of no less than thirty-two species of bipeds, and twelve of quad-rupeds. They have been observed in more than twenty localities, which are scattered over an area of nearly eighty miles from north

G

to south, in the States of Massachusetts and Connecticut.   After visiting several of these places, I entertained no doubt that the sand and mud were deposited on an area which was slowly sub-siding all the while, so that at some points a thickness of more than a thousand feet of superimposed strata had accumulated in very shallow water, the footprints being repeated at various intervals on the surface of the mud throughout the entire series of super-imposed beds."   When Sir Charles Lyell first examined this region in 1842, Professor Hitchcock had already seen two thousand impressions of feet !

It is not difficult to imagine the conditions under which such impressions may have been preserved, for at the present day there are to be seen, on some shores, illustrations of similar opera-tions.   Dr. Gould, of Boston, U.S., was the first to call the attention of naturalists to a very instructive example of such processes on the shores of the Bay of Fundy, where the tide is said to rise in some places seventy feet high.   Here we have a very perfect surface for receiving and retaining impressions. Vast are the numbers of wading and sea-birds that course to and fro over the extensive tract of plastic red surface left dry by the far retreat of the tide in the Bay of Fundy.   During the period that elapses between one spring tide and the next, the highest part of the tidal deposit is exposed long enough to receive and retain many impressions ; even during the hours of hot sunshine, to which, in the summer months, this so-trodden tract is left exposed, the layer last deposited becomes baked hard and dry, and before the returning tidal wave has power to break up the preceding one, the impressions left on that stratum have received a deposit.   A cast is thus taken of the mould previously made, and each succeeding tide brings another layer of deposit.   We can easily imagine that in succeeding ages the petrifying influences will consolidate the sandy layers into a fossil rock.   Such a rock would split in such a way, along its natural layers of formation, as to show the old moulds on one surface, and the casts on the other.

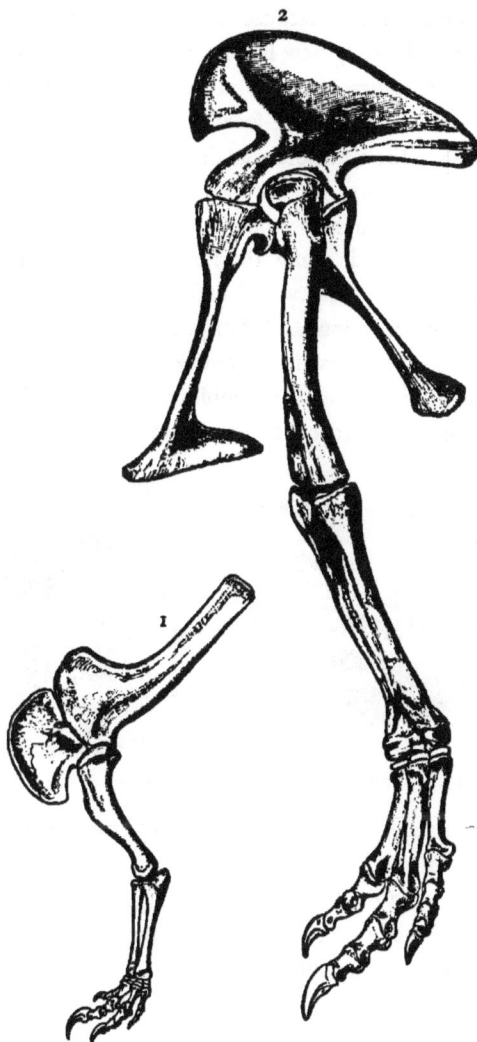

FIG. 16.—Limb-bones of *Allosaurus.* (After Marsh.)
1. Fore leg. 2. Hind leg.

Professor Marsh has had the good fortune to discover a very peculiar new form of carnivorous Dinosaur, to which he has given

the name Ceratosaurus,[1] because its skull supported a horn.   But
the horn is not the only new feature presented by this interesting
creature.   Its vertebræ are of a strange and unexpected type; and
in the pelvis all the bones are fused together, as in modern birds.
Externally, also, the Ceratosaurus differed from other members of
the carnivorous group, for its body was partly protected by long
plates in the skin, such as crocodiles have : these extended from
the back of the head, along the neck, and over the back.   An
almost complete skeleton was found which indicates an animal
about seventeen feet long.   When alive it was probably about
half the bulk of the Allosaurus mentioned above.   (See Fig. 16.)

Seen from above, its skull resembles in general outline that of a
crocodile, the facial portion being elongated and gradually taper-

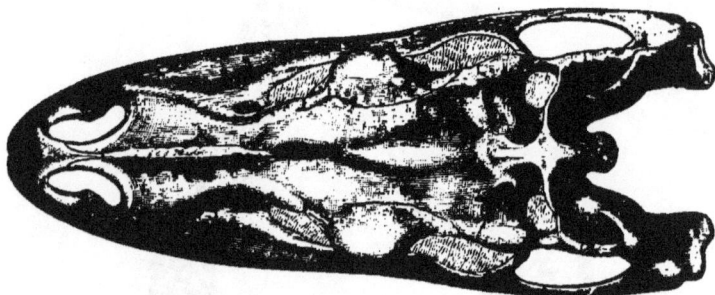

FIG. 17.—Skull of *Ceratosaurus*.   Top view.   (After Marsh.)

ing to the muzzle, with the nasal openings separate, and placed
near the end of the snout.

The teeth of this horned Dinosaur resemble those of the
Megalosaur.   Its eyes were protected by protuberances of the skull
just above the cavity in which the eye was placed (see Figs. 17 and
18).   The brain was a good deal larger in proportion to the size of
the animal than in Brontosaurus and its allies; so perhaps we may
infer that it was endowed with greater intelligence, as it certainly
was more active in its habits.   The fore limbs, as in Megalo-

[1] Greek—*keras*, horn ; *sauros*, lizard.   Some authorities consider it to be
identical with Megalosaurus.

saurus, were small, and some of the fingers ended in powerful claws, which no doubt it used to good purpose.

Perhaps the most remarkable of all the Dinosaurs was a diminutive creature only two feet in length, which was related to those we have just been considering, and whose skeleton has been found almost entire in the now famous Lithographic Stone of Solenhofen in Bavaria. Of this unique type, the Compsogna- / thus, the skeleton of which is in many ways so bird-like, Professor Huxley remarks, "It is impossible to look at the conformation of this strange reptile and to doubt that it hopped, or walked, in

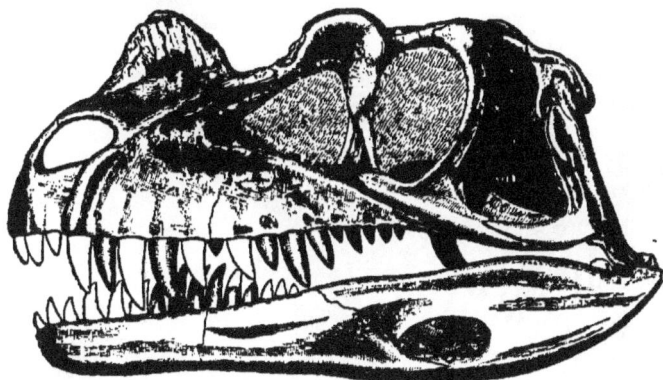

FIG. 18.—Skull of *Ceratosaurus nasicornis.* (After Marsh.)

an erect or semi-erect position, after the manner of a bird, to which its long neck, slight head, and small anterior limbs must have given it an extraordinary resemblance." (See Fig. 19.)

At the head of this chapter are placed the words of Dr. Mantell, "Fossils have been eloquently and appropriately termed *Medals of Creation,*" and the eloquent passage by which those words are followed may be transcribed here. He goes on to say, "For as an accomplished numismatist, even when the inscription of an ancient and unknown coin is illegible, can from the half-obliterated effigy, and from the style of art, determine with precision the people by whom, and the period when, it was struck : in like manner the geologist can decipher these natural memorials, interpret the hieroglyphics with which they are inscribed, and from apparently the most insignificant relics trace the history of beings of whom no other records

FIG. 19.—Skeleton of *Compsognathus longipes.* (From the Solenhofen lime-stone.)

are extant, and ascertain the forms and habits of unknown types of organisation whose races are swept from the face of the earth, ere the creation of man, and the creatures which are his contemporaries. Well might the illustrious Bergman exclaim, *" Sunt instar nummorum memoralium quæ de præteritis globi nostri fatis testantur, ubi omnia silent monumenta historica."*

Geology owes a deep debt of gratitude to the late Dr. Gideon A. Mantell, who, during the intervals of a laborious professional life, collected and described the remains of several strange extinct reptiles, and wrote a number of works on geology, such as served in his day to advance the science to which he was so enthusiastically devoted.

We propose to give a brief account of a wonderful group of Dinosaurs, first introduced to the scientific world through Dr. Mantell's labours.

The first of these monsters is the Iguanodon, the earliest known individual of the " bird-footed " division (Ornithopoda). The history of the gradual reconstruction of its skeleton is an instructive instance of the results that may be obtained by a careful and patient study of fragmentary remains. Through the labours of Dr. Mantell, in the first half of this century, a considerable knowledge was acquired of the greater part of the skeleton, but certain portions remained a puzzle; these, however, were eventually explained by Professor Huxley and Mr. Hulke, and a few years ago a series of complete skeletons were most fortunately obtained in Belgium, so that now every part of the huge framework of this monster is known to the palæontologist. Its history, as a fossil, is a most interesting one, and furnishes one more example of the marvellous insight into the nature of extinct animals displayed by the illustrious Baron Cuvier. Let us begin with the teeth, since they were the first part of the monster brought to light.

It is, perhaps, hardly necessary to remark that, to one thoroughly acquainted with the structures of living animals, a tooth, or a series of teeth, will furnish material from which important conclusions with regard to the structure and habits of an extinct animal may be drawn. So, also, with regard to some other parts, such as limb-bones, but more especially the bones of which the backbone is composed (known as vertebræ). These are very important. The veteran anatomist, Professor Owen, has said, "If I were restricted to a single specimen on which to deduce the nature of an extinct animal, I should choose a vertebra to work out a reptile, and a tooth in the case of a mammal." Seven or eight different "characters," he says, may be deduced from a reptilian vertebra. It is, of course, impossible

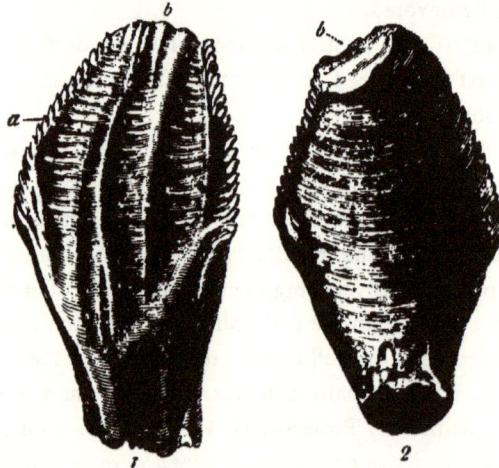

FIG. 20.—Tooth of Iguanodon, with the apex slightly worn. (From the Wealden Beds of Tilgate Forest. Natural size.) 1. Front aspect, showing the longitudinal ridges and serrated margins of the crown. 2. View of the back, or inner surface of the tooth. *a*. Serrated margins. *b*. Apex of the crown worn by use.

for any one to reconstruct an entire animal from a single bone or a few teeth, yet such fragments indicate in a general way the nature of a lost creation and its position in the animal kingdom.

It is all the more important to give to the general reader this warning, because an impression seems still to remain in the popular mind that Owen could and did restore extinct types from a single bone or a single tooth; but no anatomist would attribute to any mortal man such superhuman power. Let us, therefore, while gratefully acknowledging the debt we all owe to the great naturalist—who has gone to his rest since our first edition appeared—not attribute to him impossible things. Nor can it be denied that even he sometimes fell into error, or drew conclusions not borne out by later discoveries. It must also be confessed that in some respects he lagged behind in the march of scientific progress. While on this subject we cannot do better than quote some remarks of our friend, Mr. A. Smith Woodward, of the Natural History Museum, in an able review of Sir Richard's work on vertebrates.[1] He says, "Owen, in fact, was Cuvier's direct successor, and, apart from his striking hypotheses . . ., it is in this character that he has left the deepest impression upon biological science. Extending and elaborating comparative anatomy as understood by Cuvier, Owen concentrated his efforts on utilising the results for the interpretation of the fossil remains —even isolated bones and teeth—of extinct animals. He never hesitated to deal with the most fragmentary evidence, having complete faith in the principles established by Cuvier; and it is particularly interesting, in the light of present knowledge, to study the long series of successes and failures that characterise his work. However, unwittingly, Owen may be said to have contributed most to the demolition of the narrow Cuvierian views. When dealing with animals closely related to those now living, his correctness of interpretation was usually assured; when treating of more remote types, he could do little more than guess, unless tolerably complete skeletons happened to be at his disposal. . . .

"In short, Owen's work on fragmentary fossils has demonstrated that the principles of comparative anatomy are very different from those inferred by Cuvier from his limited field of observation, and the discoveries of Leidy, Marsh, Cope, Scott, and Osborn, in America, have finally led to a new era that Owen only began to foresee clearly in his later days."

The first specimens of teeth of the Iguanodon were found by

[1] *Natural Science*, ii. p. 130. (Feb. 1893.)

Mrs. Mantell, in 1822, in the coarse conglomerate of certain strata in Tilgate Forest, belonging to the Cretaceous period (see Table of Strata, Appendix I.). Dr. and Mrs. Mantell subsequently collected a most interesting series of these remarkable teeth (which, for a time, puzzled the most learned men of the day), from the perfect tooth of a young animal, to the last stage, that of a mere long stump worn away by mastication. In external form they bore a striking resemblance to the grinders of herbivorous mammals, and were wholly unlike any that had previously been known. Even the quarrymen, accustomed to collect the remains of fishes, shells, and other objects embedded in the rocks, had not observed fossils of this kind; and until Dr. Mantell showed them his specimens, were not aware of the presence of such teeth in the stone they were constantly breaking up for the roads. The first specimen that arrested his attention was a large tooth, which, from the worn surface of its crown, had evidently once belonged to some herbivorous animal. In form it so entirely resembled the corresponding part of an incisor tooth of a large pachydermatous animal ground down by use, that Dr. Mantell was much embarrassed to account for its presence in the ancient Wealden strata, in which, according to all previous experience, no fossil remains of mammals would be likely to occur. No reptiles of the present day are capable of masticating their food; how, then, could he venture to assign it to a reptile? Here was a puzzle to be solved, and in his perplexity he determined to try whether the great naturalist at Paris would be able to throw any light on the question. Through Sir Charles (then Mr.) Lyell, this perplexing tooth was submitted to Baron Cuvier; and great was the doctor's astonishment on hearing that it had been without hesitation pronounced to be the upper incisor of a rhinoceros! The same tooth, with some other specimens, had already been exhibited at a meeting of the Geological Society, and shown to Dr. Buckland, Mr. Conybeare, and others, but with no more satisfactory result. Worse than that: Dr. Mantell was told that the teeth were of no particular interest, and that, without doubt, they either belonged to some large fish, or were the teeth of a mammal, and derived from some superficial deposit of the "glacial drift," then called Diluvium.

There was one man, however, who foresaw the importance of Mantell's discovery, and that was Dr. Wollaston. This distinguished philosopher, though not a naturalist, supported the doctor's idea that the teeth belonged to an unknown herbivorous reptile, and encouraged him to continue his researches.

As if to add to the difficulty of solving the enigma, certain bones of the fore limb, discovered soon after in the same quarry and forwarded to Paris, were declared to belong to a species of hippopotamus! Another very curious bone—of which we shall speak presently—was declared to be the lesser horn of a rhinoceros! The famous Dr. Buckland even went so far as to warn Dr. Mantell not to publish it forth that these bones and teeth had been found in the Tilgate Forest strata. To him it seemed incredible that such remains could have been obtained from beds older than the superficial drift deposits of the district. We must bear in mind that in those days palæontology, or the knowledge of the world's former inhabitants, was a new science still in its infancy, and the idea of mammals having existed so far back as the Cretaceous period must have appeared incredible.

However, the workmen in the quarry were stimulated by suitable rewards, and at length the doctor's efforts resulted in the discovery of teeth which displayed the curious serrated edges, and the entire form of the unused crown. Having forwarded specimens and drawings of these to Paris, Dr. Mantell went to London, and ransacked all the drawers in the Hunterian Museum that contained jaws and teeth of reptiles, but without finding any that threw light on this subject. Fortunately, Mr. Samuel Stuchbury, then a young man, was present, and proposed to show him the skeleton of an Iguana, which he had himself prepared from a specimen that had long been immersed in spirits.

And now the puzzle was in a fair way to being solved; for, to his great delight, the doctor found that the minute teeth of that reptile bore a closer resemblance in their general form to those from Tilgate Forest than any others he had ever seen.

In spite of this fortunate discovery, however, others remained obstinate and unconvinced; and it was not until he had collected a series of specimens, exhibiting various stages of the teeth, that the correctness of his opinion was admitted, either as to their true interpretation, or the age of the strata in which they were imbedded. And now there came good news from Paris. Cuvier, with the fresh material submitted to him, had boldly renounced his previous opinion, and gave the weight of his great authority to the view maintained by the discoverer of these teeth. In a letter to the doctor he said that such teeth were quite unknown to him, and that they belonged to some reptile. He suggested that they implied the existence of a *new animal*, a *herbivorous reptile*. Time would either confirm or disprove the idea, and in the mean time he advised Dr. Mantell to seek diligently for further evidence, and, if part of a jaw could be found with teeth adhering, he believed he could solve the problem. In his immortal work, *Ossemens Fossiles*, Cuvier generously admits his former mistake, and said he was entirely convinced of his error.

Baron Cuvier alone amongst the doctor's friends or correspondents was able to give any hint as to the character and probable relations of the animal to which the recently discovered teeth belonged. Being hampered by arduous professional duties in a provincial town, remote from museums and libraries, Dr. Mantell transmitted to the Royal Society figures and drawings of the specimens, and, at the suggestion of the Rev. W. D. Conybeare, adopted the name Iguanodon (Iguana-tooth) for the extinct reptile, a name which pointed to the resemblance of its teeth to those of the modern iguana, a land-lizard inhabiting many parts of America and the West Indies, and rarely met with north or south of the tropics. These lizards are from three to five feet in

length, and perfectly harmless, feeding on insects and vegetables,
and climbing trees in quest of the tender leaves and buds, which
they chip off and swallow whole; they nestle in the hollows of
rocks, and deposit their eggs in the sands and banks of rivers.

In all living reptiles the insects or vegetables on which they
feed are seized by the tongue or teeth, and swallowed whole, so
that a movable covering to the jaws, similar to the lips and
cheeks of the mammalia, is not necessary, either for seizing and
retaining food, or for subjecting it by muscular movements to the
action of the teeth. It is the power of perfect mastication
possessed by the Iguanodon that is so strange, for it implies a
most remarkable approach in extinct reptiles to characters possessed now only by herbivorous mammalia, such as horses, cows,
deer, etc. From this and other strange characters seen in the
Dinosaurs, we learn that they in their day played the part of our
modern quadrupeds, whether carnivorous or herbivorous, and
showed a remarkable approach to the mammalian type, which of
course is a much higher one.

It is, therefore, not to be wondered at that Dr. Mantell's contemporaries, with the exception of Cuvier, found in the teeth we
have described an awkward puzzle, and refused to believe that
they belonged to a reptile. Such a notion was at variance with
all previous experience, and we naturally form our conclusions to
a large extent by experience. Let us, then, beware lest we allow
our ideas to be limited by what after all is, as it were, only an
expression of our ignorance. The Hottentot who has never seen
snow would refuse to believe that rain can assume a solid form;
and, in the same way, if we bind ourselves down by experience,
we might refuse to believe in some of the still more wonderful
dinosaurian types to be described in this chapter, such as the
Triceratops, with a pair of large horns, a skull over six feet
long, and limbs larger than those of the rhinoceros! (see p.
117).

The strange vagaries of Dinosaurs have led Professor Marsh

and other authorities to exalt them, from their former position of a mere order in the reptile class, to the dignity of a sub-class all to themselves; and there is much to be said for this view. Compared with the Marsupials, living and extinct, they show an equal diversity of structure and variations in size from by far the largest land animals known down to some of the smallest.[1]

The importance of discovering, if possible, a portion of the jaw of an Iguanodon was fully recognised by Dr. Mantell, and, urged on by the encouragement he had received from the illustrious Cuvier, he eagerly sought for the required evidence. But nearly a quarter of a century elapsed before it was forthcoming. In 1841 and 1848, however, portions of the lower jaw, with some teeth attached, were found; and his memoir *On the Structure of the Jaws and Teeth of the Iguanodon* was published by the Royal Society in 1848. For this important communication the gold medal of the society was awarded to the author. The second of these finds (by Captain Brickenden) confirmed in every essential particular the inferences suggested by the detached teeth.

The first important connected series of bones of this monster was discovered in 1834, by Mr. Bensted, in the "Kentish Rag" quarries of the Lower Greensand formation at Maidstone. Mr. Bensted, who was the proprietor of the quarry, one day had his attention drawn by the workmen to what they supposed to be petrified wood in some pieces of stone which they had been blasting. He perceived that what they supposed to be wood was fossil bone, and, with a zeal and care which have always characterised this estimable man (says Professor Owen) in his endeavour to secure for science any evidence of fossil remains in his quarry, he immediately resorted to the spot. He found that the bore, or blast, by which these remains were brought to light

---

[1] Bauer, after a full critical examination of the Dinosauria, considers that one order is insufficient, and has proposed to make three orders of them, which he names after the Iguanodon, Cetiosaurus, and Megalosaurus.

had been inserted into the centre of the specimen, so that the mass of stone containing it had been shattered into many pieces, some of which were blown into the adjoining fields! All these pieces he had carefully collected, and, proceeding with equal ardour and success to the removal of the matrix from the fossils, he succeeded, after a month's labour, in exposing them to view, and in fitting the fragments in their proper place. This valuable specimen was presented to Dr. Mantell (and afterwards purchased with the rest of his collection by the British Museum), and its present condition is the result of his skill, as well as that of its discoverer. Certain gentlemen in Brighton, anxious that the specimen should be placed in the hands of the original discoverer of Iguanodon, purchased and presented it to Dr. Mantell—a tribute of respect which was highly gratifying to him. (Wall-case 6.)

It belonged to a young Iguanodon. This fortunate discovery was one of those Cuvier foresaw, and has served to verify his sagacious conjecture that some of the great bones collected by the doctor from the Wealden strata of Sussex belonged to the same animal, and to confirm other conclusions formed by the discoverer of the Iguanodon. Great was Dr. Mantell's delight on finding that every bone he had ascribed to Iguanodon solely from analogy was present in the Maidstone specimen. One of the chief advantages of this discovery was that it afforded demonstration of the characters of the vertebræ, which, as previously stated, are very important to the anatomist. Of these Professor Owen has given full descriptions, and has shown that they differ from those of any animal previously known, whether living or extinct.

It is very interesting, in the light of recent discoveries, to read the conclusions arrived at by Mantell and Owen, with regard to the organisation of this great Wealden reptile, and to see how, with the exception of certain details, they have been confirmed. Considering the imperfect nature of the materials at their command, it is wonderful that their forecasts should have turned out so successful. Thus Professor Owen predicted for the Iguanodon a total length of twenty-eight feet, and specimens discovered of late years show a length of twenty-four feet. In some, the thigh-bone exceeded a yard in length; this indicated an animal of great size, since in the largest crocodiles this bone is scarcely a foot long. Again, Dr. Mantell, from a study of the imperfect jaw-bones in his collection, concluded that the lower jaw was invested with a well-developed fleshy flexible lip, and that the mouth was provided with a tongue of great mobility and power. "There are strong reasons," he says, "for supposing that the lip was flexible, and, in conjunction with the long fleshy prehensile tongue, constituted the instrument for seizing and cropping the leaves and branches, which, from the construction of the molars, we may infer, constituted the chief food of the Iguanodon. The mechanism of the maxillary organs (jaws), as elucidated by recent discoveries, is thus in perfect harmony with the remarkable characters which rendered the first known teeth so enigmatical; and in the Wealden herbivorous reptile we have a solution of the problem, how the integrity of the type of organisation peculiar to the class of cold-blooded vertebrata was maintained, and yet adapted, by simple modifications, to fulfil the conditions required by the economy of a gigantic terrestrial reptile, destined to obtain support exclusively from vegetable substances; in like manner, as the extinct colossal herbivorous Edentata (sloths, see Chapter XII.), which flourished in South America ages after the country of the Iguanodon and its inhabitants had been swept away from the face of the earth."

Dr. Mantell also was the first to prove, from the nature of the Wealden strata, that they were deposited in or near the estuary of a mighty river. With regard to the aspect of the country in which the Iguanodon flourished, he showed that coniferous trees probably clothed its Alpine regions; palms and arborescent ferns, and cycadaceous plants (*i.e.* plants resembling the modern

A GIGANTIC DINOSAUR, IGUANODON BERNISSARTENSIS.
Length about 30 feet.

PLATE VII.

zamia, or " false palm "), constituted the groves and forests of its plains and valleys ; and in its fens and marshes the equisetaceæ (mare's-tails) and plants of a like nature prevailed.

The Iguanodons of the Wealden epoch did not live and die where their bones are now found—the condition in which their fossil relics occur proves that they floated down the streams and rivers, with rafts of trees and other spoils of the land, till, arrested in their course, they sank down and became buried in the fluviatile and sometimes marine sediments then being slowly laid down. In this way only can we account for the generally broken and rolled condition of the bones, their separation from each other, the numerous specimens of teeth which must have been detached from their sockets, and the broken stems and branches of trees without leaves that have been found in the Wealden strata of England.

Since the days of Dr. Mantell, the remains of Iguanodon, or closely allied genera, have been found on the continent, in other parts of England, and in North America, in strata of various ages, from the Trias or New Red Sandstone to the Chalk (see Table of Strata, Appendix I.). The American Hadrosaurus must have decidedly resembled the Iguanodon.

The beautiful restoration by our artist (plate VII.) is based upon the Belgian specimens described in the following chapter.

CHAPTER VII.

DINOSAURS (*continued*).

"Everything in Nature is engaged in writing its own history: the planet and the pebble are attended by their shadows, the rolling rock leaves its furrows on the mountain side, the river its channel in the soil, the animal its bones in the stratum, the fern and the leaf inscribe their modest epitaphs on the coal, the falling drop sculptures its story on the sand and on the stone,— not a footstep on the snow or on the ground, but traces in characters more or less enduring the record of its progress."—EMERSON.

IN the year 1878 was announced one of the most fortunate discoveries known in the whole history of geological science —a discovery unique of its kind, and one which throws considerable light on the nature of the monster first discovered by Dr. Mantell. In that year came the good news that no less than twenty-three Iguanodons had been found in the colliery of Bernissart, in Belgium, between Mons and Tournai, near the the French frontier. The coal-bearing rocks (coal-measures) of this colliery, overlain by chalk and other deposits of later age, are fissured in many places by deep valleys or chasms more than 218 yards deep. Though now filled up, they must at one time have been open gorges on an old land surface. Into one of these chasms were somehow precipitated twenty-three Iguanodons, numbers of fish, a frog-like animal, several species of turtles, crocodiles, and numerous ferns similar to those described by Mantell from the Weald. It it not easy to conjecture how this large and varied assemblage of animals came to be collected together and entombed in this one place, but possibly their carcases were swept by some flood into the chasm in which the remains were discovered. They were buried in clay interstratified with sand, a fact which was interpreted in accordance with the above suggestion.

M. de Pauw, the accomplished controller of the workshops in the Royal Museum of Natural History at Brussels, spent three whole years in extracting this splendid series of fossils from the pit-shaft, the bones being brought up from a depth of rather more than 350 yards. But at the end of this time it was only the rough material that had been got together, and every block containing bones requires a great deal of most careful labour before the bones in it are so exposed that they can be properly studied. Out of the twenty-three specimens, fifteen had, in the year 1883, been chiselled out, eight remaining to be worked at; and although five skilled workmen were then constantly at work, progress was necessarily slow.

In 1883, that is after seven years, two huge entire skeletons had been set up in a great glass case in the Courtyard of the Museum at Brussels, and these exhibit with marvellous completeness the structure of the extinct monster.[1] The work reflects the highest credit on M. de Pauw;[2] and the director of the Bernissart Mining Company, M. Fages, deserves the thanks of all scientific men for so liberally aiding this important undertaking. These specimens illustrate the conclusion, previously arrived at by Professor Huxley, that Dinosaurs, as a group, occupy a position in the great chain of animal life intermediate between reptiles and birds. Indeed, it is the opinion of this great authority, and of many naturalists of the present day, that whenever future discoveries may reveal the ancestry of birds, it will be found that they came from Dinosaurs, or that both originated from a common ancestor.

The specimens so skilfully set up by M. de Pauw represent

[1] In August, 1892, Mr. Dollo wrote, in answer to inquiries from South Kensington, to say that five are already mounted and exhibited, and five more are almost ready for mounting. He also stated that the remains represent twenty-nine individuals, not twenty-three, as above.

[2] *Geological Magazine*, January, 1885.

two distinct species. The larger one, Iguanodon Bernissartensis, cannot be less than fifteen feet high, and, measured from the tip of the snout to the end of the tail, is rather over thirty feet long, covering nearly twenty-four feet of ground in its erect position (see Fig. 21). Iguanodon Mantelli is smaller and more slender looking, with a height of over ten feet, and a length of about twenty feet. (See Fig. 22.)

The huge three-toed impressions found in Sussex prove that

FIG. 21.—Skeleton of *Iguanodon Bernissartensis.*

the monster, although owning a body as large as that of an elephant, habitually walked on its hind legs! Some of the thigh-bones found by Dr. Mantell measured between four and five feet in length. It will be seen that the fore limbs are small in com-parison to the hind limbs. A remarkable feature of the hand is the large pointed bone at the end of the thumb, forming a kind of spur. The conical shape of this bone found by Dr. Mantell, who

IGUANODON MANTELLI.
Length about 20 feet.

PLATE VIII.

had no clue to its place in the skeleton, led him to suppose that it was a horn answering to that of a rhinoceros—a conclusion which Professor Owen refused for various reasons to accept. The latter concluded that it belonged to the hand, and now we see that he was right. Unfortunately, certain popular works on geology, such as *Our Earth and its Story* (Cassell) still continue to spread this error, by showing a (very indifferent) restoration of

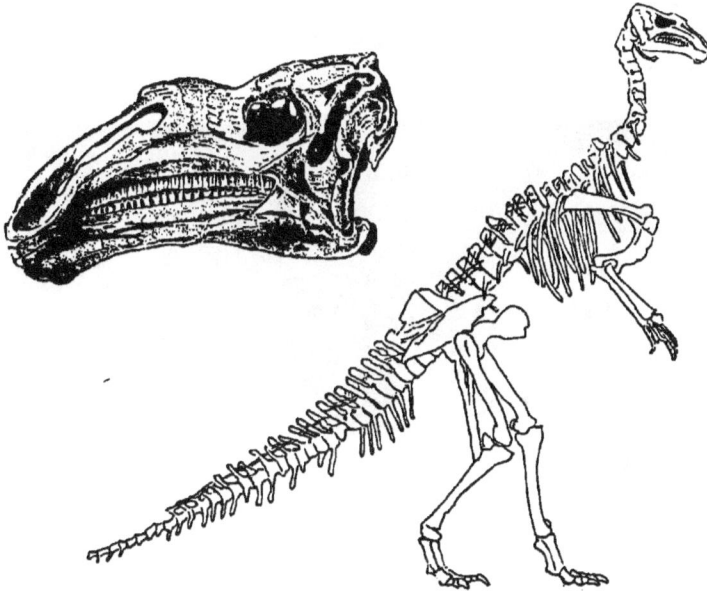

Fig. 22.—Skull and skeleton of *Iguanodon Mantelli.* (From Bernissart.)

the Iguanodon with the impossible horn on its nose. It has been suggested that the spur was a weapon of offence, and that, when attacked, an Iguanodon may have seized its aggressor in its short arms, and made use of the spur as a dagger. But this is only conjecture, and perhaps the spur may have been useful in seizing and pulling down the foliage and branches of trees, or in grubbing them up by the roots. Detached specimens of this curious bone may be seen among the other remains of Iguanodon

at South Kensington, and also some of the gigantic tracks already
alluded to. (Gallery IV. on plan, Wall-cases 5 and 6; and
Gallery XI., Wall-case 7.)

The Bernissart specimens even afford some evidence as to
the nature of the integument, or skin, and this supports the
idea previously held that the creature possessed a smooth skin,
or, at least, only slightly roughened. The muzzle was quite toothless, and perhaps may have been sheathed in horn, like the beak
of turtles—an arrangement highly useful for biting off the leaves of
trees.

FIG. 23.—Tracks of *Iguanodon*, much reduced. (From Wealden strata,
Sussex.)

Probably it passed much of its time in the water, using its
immense powerful tail as an organ of propulsion. When
swimming slowly it may have used both sets of limbs, but when
going fast it probably fixed its fore limbs closely beside its body,
and drove itself through the water by means of the long hind
limbs alone. Mr. Dollo, of Brussels, is preparing a final monograph on the Bernissart Iguanodons, a work to which palæontologists eagerly look forward. There cannot be much doubt that

these unarmoured Dinosaurs were molested and preyed upon by their carnivorous contemporaries, such as the fierce Megalosaurus, previously described (p. 76). And with regard to this, Mr. Dollo makes the suggestion that, when on land, their great height and erect posture enabled them to descry such enemies a long way off. Their great height must also have stood them in good stead, by enabling them easily to reach the leaves of trees, tree-ferns, cycads, and other forms of vegetable life, which constituted their daily food. (See restorations, Plates VII. and VIII.)

Should the reader visit the " geological island " in the grounds of the Crystal Palace, he will see that Mr. Waterhouse Hawkins's great model Iguanodon there set up is by no means in accordance with the description given above; but we must remember how imperfect was the material at his command.

Another Dinosaur, of considerable dimensions, that flourished during the Wealden period was the Hylæosaurus, also discovered by Dr. Mantell, and so named by him because it came from the Weald.[1] In the summer of 1832, upon visiting a quarry in Tilgate Forest, which had yielded many organic remains, he perceived in some fragments of a large mass of stone which had recently been broken up and thrown in the roadside, traces of numerous pieces of bone. With great care he cemented together and fixed in a stout frame, all the portions of this block that he could find, and set to work to "develop" the block with his chisel. This work occupied many weeks, but his labour was rewarded by the discovery of certain new and remarkable features displayed by this monster; for it must have presented, when alive, a formidable array of bony plates and long sharp spines, the latter of which probably stood in bristling array along the back and tail, and other parts of the body. (Wall-case 4.) Of the spines no less than ten were found in this block, varying in length from five to seventeen inches, the largest being four inches thick. It is known that many lizards, such as Iguanas

---

[1] From Greek—*hule*, wood, or weald; and *sauros*, lizard.

and Cycluras, have large processes with horny coverings, forming a kind of fringe or crest along the back, and, judging by analogy, Dr. Mantell concluded that this gigantic saurian was similarly armed with a row of large angular spines covered by a thick horny investment. As weapons of offence and defence, they were no doubt highly effective, but their precise arrangement is still a matter of speculation.

This first specimen displayed, besides the bony scutes and spines, a portion of the backbone, eleven ribs and portions of the pectoral arch. A second specimen was found near Bolney, in Sussèx, and was unfortunately almost wholly destroyed by the labourers ; but Dr. Mantell was able to obtain many of the bones, such as ribs and limb-bones, and they also indicated a reptile of great size. A third specimen was brought to light in Tilgate Forest in 1837 ; but, unfortunately, this also fell into the hands of the parish labourers, who were unacquainted with its value. Although with due care a much larger portion of the skeleton might have been kept, yet Dr. Mantell was able to obtain a fine series of twenty-six vertebræ belonging to the tail, with a total length of nearly six feet : the same spines were present here also.

No specimen of the skull of this strange monster is known, and no teeth that can be with certainty referred to it.

Mr. Waterhouse Hawkins's model at Sydenham, near the Iguanodon, was based on the above discoveries, which are insufficient, and is far from the truth.

The next monster to be described is one that has fortunately left to posterity a much better record of itself, and probably was not very unlike the Hylæosaurus of Mantell. This is the Scelidosaurus : so named by Professor Owen from the indications of greater power in the hind legs than in most saurians.[1] It is the only known example of an almost entire skeleton of an English Dinosaur, and the history of its discovery is rather

---

[1] From Greek—*scelis*, limb, and *sauros*, lizard.

PLATE IX.

AN ARMOURED DINOSAUR, SCELIDOSAURUS HARRISONI.
Length 12 feet or more.

curious. Some time previous to 1861, Mr. J. Harrison, of
Charmouth, obtained from the Lower Lias of that neighbourhood
portions of the hind limb of a Dinosaur, and, later on, a nearly
complete skull. These specimens were described by Owen, and
the genus was founded on them. Mr. Harrison, whose discovery
aroused great interest, continued to search on the same spot,
and was rewarded by finding all the rest of the skeleton, except

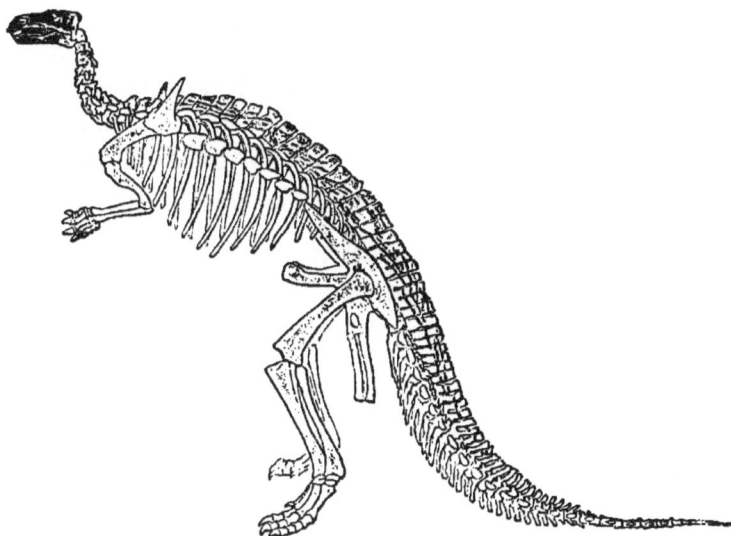

FIG. 24.—Restored skeleton of *Scelidosaurus Harrisoni* (after Woodward),
greatly reduced, from the Lower Lias of Charmouth, Dorset. The figure
shows the large lateral dermal spines on the shoulders, and the long lateral
line of smaller spines, reaching from the pectoral region to the extremity of the
tail.

most of the neck vertebræ. This was extracted in several blocks,
and these, after careful "development" of the bones, were fitted
together so as to exhibit the whole skeleton. This most valuable
specimen can now be seen at South Kensington in a separate glass
case, and is one of the treasures of the unrivalled gallery of fossil
reptiles. The case is placed so that both sides of the specimen
can be seen (Case Y, Gallery IV., on plan). Its length is about

twelve feet ; perhaps the individual it represents was not fully grown, but, on account of the absence of most of the neck vertebræ, it is impossible to give the exact length. Both hind limbs are entire and well seen, but of the fore limbs the hands are wanting. The former were provided with four "functional" toes—that is, toes that were used,—and one "rudimentary" or unused one. There were two big spines, one placed on each shoulder, and a series of long plates arranged in lines along the back and side. Plate IX. shows an attempted restoration of this remarkable Dinosaur based upon the skeleton just described. It seems to have been organised for a terrestrial rather than an aquatic life, but to have been amphibious, frequenting the margins of rivers or lakes. Professor Owen considers that the carcase of this individual drifted down a river emptying itself in the old Liassic Sea, on the muddy bottom of which it would settle down when the skin had been so far decomposed as to permit the escape of gases due to decomposition. In that case the carcase would attract large carnivorous fishes and reptiles, such as swarmed in this old sea, so that portions of the skin and flesh would probably be torn away before the weight of the bones had completely buried it in mud. In this way, perhaps, the loss of much of the external armature and of the two fore feet may be accounted for. The hind limbs, being stronger, were better able to resist such attacks, and they are therefore preserved. Like many other specimens, this fossil has, in the course of ages, been subjected to enormous pressure from overlying strata, causing compression and dislocation or fracture.

But there were in existence during the long Jurassic period, other and even stranger forms of armoured Dinosaurs. One of these, only imperfectly known at present, was the many-spined Polacanthus.[1] This remarkable monster had the whole region of the loins and haunches protected by a continuous sheet of bony plate armour, rising into knobs and spines, after the fashion of the

[1] From Greek—*polus*, many, and *acantha*, spine.

shield or carapace of certain extinct armadillos known as Glyptodonts (see Chapter XII.). A specimen of such a shield is to be seen in the collection at South Kensington (Wall-case 4). It is to be hoped that, some day, further remains of the Polacanthus will be brought to light, so that a restoration may become possible. Dr. Mantell had already pointed out certain analogies between Iguanodon and the huge extinct sloths of the South American continent, that flourished in the much more recent Pleistocene period; and this idea is now considerably strengthened by the later discoveries of armoured Dinosaurs. These are his words : " In fine, we have in the Iguanodon the type of the terrestrial herbivora which, in the remote epoch of the earth's physical history termed by geologists *the age of Reptiles*, occupied the same relative position in the scale of being, and fulfilled the same general purposes in the economy of nature, as the Mastodons, Mammoths, and Mylodons (extinct sloths) of the Tertiary period, and the existing pachyderms."

It is, perhaps, one of the most interesting discoveries of modern geology, that certain races of animals now extinct have in various ways assumed some of the characteristics presented by animals much higher in the scale of being, that flourish in the present day. It seems as if there had been some strange law of anticipation at work, if we may venture so to formulate the idea. It has already been shown how the great saurians Ichthyosaurus and Plesiosaurus presumed to put on some of the characters of whales, and to play their *rôle* in nature, though they were only reptiles ; how the carnivorous Dinosaurs acquired teeth like those now possessed by lions and tigers, which also are mammals ; and now we find herbivorous Dinosaurs imitating the Glyptodon, an armadillo that lived in South America almost down to the human period. We shall not lose sight of this very interesting and curious discovery, for other cases will present themselves to our view in future chapters. The reader might ask, " If reptiles were able in these and other ways to imitate the mammals of to-day, or of yesterday, why should they not have been able to go a few steps further, and actually *become* mammals ? "   The Evolutionist, if confronted with such a question, would say, that there is no evidence of Dinosaurs turning into mammals, but that both may have branched off at an early geological period (say the Permian) from a primitive group of reptiles, or even of amphibians.

It must be borne in mind that, during the "age of reptiles" (Mesozoic period), the mammalian type was but feebly represented by certain small and humble forms, probably marsupials.   As far as we know, there were no big quadrupeds such as flourish to-day ; therefore reptiles played their part, and in so doing acquired some of their habits and structural peculiarities.   It is difficult for us, living in an age of quadrupeds, to realise this, and to picture to ourselves reptilian types posing as "lords of creation," or, to use a homely phrase, "strutting in peacock's feathers."

Leaving now the English herbivorous Dinosaurs, we pass on to those still more wonderful forms discovered of late years by Professor Marsh.   The former have been treated at considerable length, first because they are English, and, as such, the history of their discovery possesses considerable interest ; secondly, because their elucidation reflects the highest credit on our great pioneers in this fruitful field of research, and illustrates the manner in which great naturalists have been able to draw most important and wonderful conclusions (afterwards verified in most cases) from material apparently far from promising.   For example, Cuvier's prophecy of the Iguanodon from a few teeth is a striking example of the result of reasoning from the known to the unknown, an example which seems to us worthy to be ranked with the discovery of Neptune by Adams and Leverrier, or, to take a more recent case, the discovery by Mendeleef of the Periodic Law, by means of which he has foretold the discovery of new chemical elements.

Whatever may have been the origin of the great mammalian

class, the possibility and even probability of birds and Dinosaurs being descended from a common ancestor is a theory for which much may be said, and it has been adopted by many leading naturalists of the present day, who have been convinced by Professor Huxley's clear elucidation of the nature of the pelvic region in the group of Dinosaurs which has been above described (the Ornithopoda, or bird-footed group). It was Professor Huxley who first propounded this interesting speculation, basing his belief on the many bird-like characters presented by this strange group of extinct reptiles—the small head and fore limbs, the long and often three-toed hollow hind limbs, the bones of the pelvis or haunch, their habit of walking in a semi-erect position on those limbs (as proved by their tracks), and in some of hopping, as the little Compsgnathus most probably did. And, last but not least, the strange mixture of bird-like and reptilian characters presented by certain most anomalous birds discovered by Professor Marsh in American Cretaceous rocks, viz. the huge Hesperornis and the smaller Ichthyornis. Speaking on this subject some years ago, Professor Marsh said, " It is now generally admitted by biologists who have made a study of vertebrates, that birds have come down to us through the Dinosaurs, and the close affinity of the latter with recent struthious birds (ostrich, etc.), will hardly be questioned. The case amounts almost to a demonstration, if we compare with Dinosaurs their contemporaries, the Mesozoic birds. The classes of birds and reptiles, as now living, are separated by a gulf so profound that a few years since it was cited by the opponents of Evolution as the most important break in the animal series, and one which that doctrine could not bridge over. Since then, as Professor Huxley has clearly shown, this gap has been virtually filled by the discovery of bird-like reptiles and reptilian birds. Compsognathus and Archæopteryx of the Old World, and Ichthyornis and Hesperornis of the New, are the stepping-stones by which the

Evolutionist of to-day leads the doubting brother across the shallow remnant of the gulf, once thought impassable." [1]

We now pass on to describe two of the strangest and most wonderful of all the Dinosaurs, recently discovered in the far West. The first of these is the Stegosaurus,[2] or plated lizard, not wholly unknown before, because part of its skeleton was found some years ago in a brickfield in the Kimmeridge Clay at Swindon. It has been proved that some of the bones to which the name Omosaurus [3] has been applied really belonged to the former genus.

With such complete specimens now known by Professor Marsh's descriptions, it will not be necessary to mention the meagre remains discovered in this country, or the conclusions arrived at by Owen and Seeley, interesting as they are.

In the year 1877 Professor Marsh described, in the *American Journal of Science*, a considerable portion of a skeleton of a Stegosaur, remarking that this genus proved to be one of the most remarkable animals yet discovered. It was found on the eastern flank of the Rocky Mountains, in strata of Jurassic age; they indicated an animal about twenty-five feet long, and for this discovery Science is indebted to Professor A. Lakes and Engineer H. C. Beckwith of the United States Navy, who found the remains in Colorado, near the locality of the gigantic Atlantosaurus. The solid limb-bones seem to point to an aquatic life, but there can be little doubt that the monster did not pass all its time in the water. (Fig. 25 shows the skeleton.) [4]

In 1879 Professor Marsh announced the discovery of additional

---

[1] *The Introduction and Succession of Vertebrate Life in America.* An address delivered before the American Association for the Advancement of Science, at Nashville, Tenn., August, 1877. See *Nature*, vol. xvi.

[2] Greek—*stegos*, roof or covering ; *sauros*, lizard.

[3] Greek—*omos*, humerus, and *sauros*, lizard.

[4] The writer is informed that this skeleton is not yet mounted in the Yale College Museum, but that it will be before long. Our artist has drawn it as if set up, with a man standing by for comparison.

remains from several localities. The most striking feature—from which the Stegosaur takes its name—was the presence of huge bony plates belonging to its skin, as well as large and small spines. Some of the plates were from two to three feet in diameter, and they were of various shapes. Of the spines, some were of great size and power, one pair being each over two feet long! The skull was remarkably small, and more like that of a lizard than we find in most Dinosaurs; the jaws were short and massive. Little was known at first of the brain, but fortunately a later discovery showed the brain-case well preserved. Later still, more than twenty other specimens of this Dinosaur were obtained, so that nearly every portion of the skeleton is now known. The skulls indicate that the creature possessed large eyes and a considerable power of smell. The jaws contain but a single row of teeth in actual use; but as these wore out, they were replaced by others lodged in a cavity below. Teeth, however, were not its strong point; they indicate a diet of soft succulent vegetation. The vertebræ have the faces of their centra more or less bi-concave. Many curious features in the skeleton can only be explained with reference to the heavy armour of plates and spines with which the Stegosaur was provided. Thus the vertebræ have their "neural spines" expanded at the summit to aid in supporting part of the armour. (See Fig. 26.) The fore limbs were short and massive, but provided with five fingers; the hind limbs were very much larger and more powerful. These and the powerful tail show that the monster could support itself on them as on a tripod, in an upright position, and this position must have been easily assumed in consequence of the massive hind quarters. As in Iguanodon, there were three toes to the hind feet, and these were probably covered by strong hoofs. The fore limbs could move freely in various directions like a human arm, and were probably used in self-defence. (See Fig. 27.) But for this purpose the tail with its four pairs of huge spines would be very effective, and one could

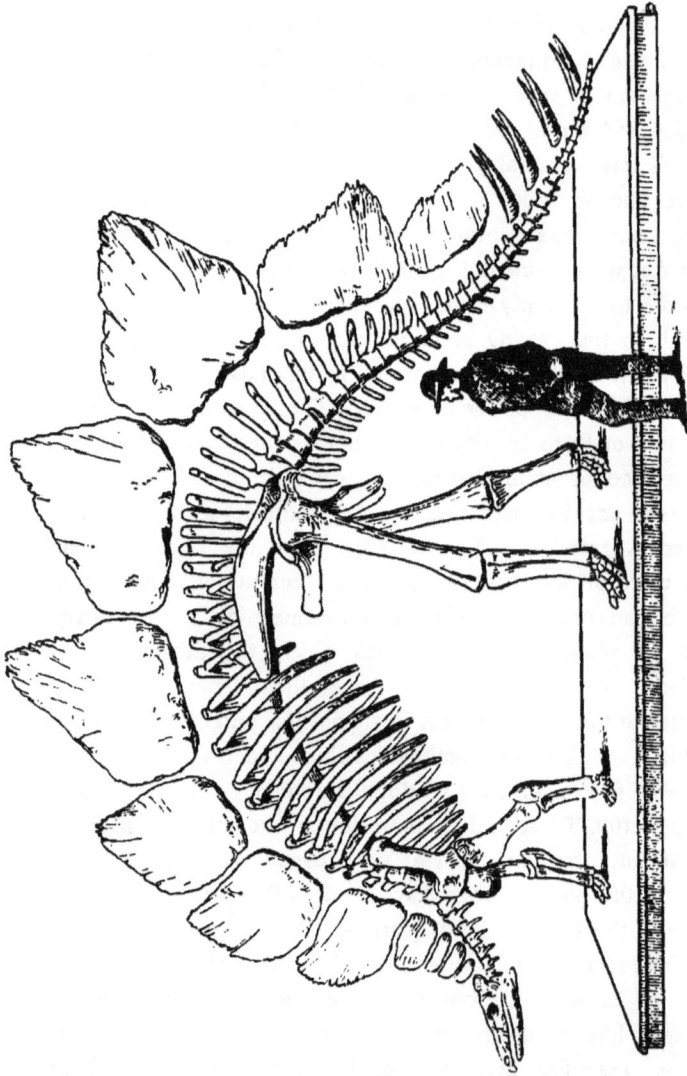

FIG. 25.—Skeleton of *Stegosaurus ungulatus* ; length about 25 feet.   (After Marsh.)

A GIGANTIC ARMOURED DINOSAUR, STEGOSAURUS UNGULATUS.
Length about 30 feet.

PLATE X.

easily imagine that a single deadly blow from such a tail would be sufficient to drive away, if not to kill, one of the carnivorous enemies of the species. All the plates and spines were, during life, protected by a thick horny covering, which must have increased their size and weight. Such a covering seems to be clearly indicated by certain grooves and impressions that mark their surfaces. (See Fig. 28.) The largest plates are unsymmetrical, and were probably arranged along the back, as in our restoration, Plate IX. It will be noticed, by those who are familiar with our first edition, that Plate X. gives a somewhat different representa-

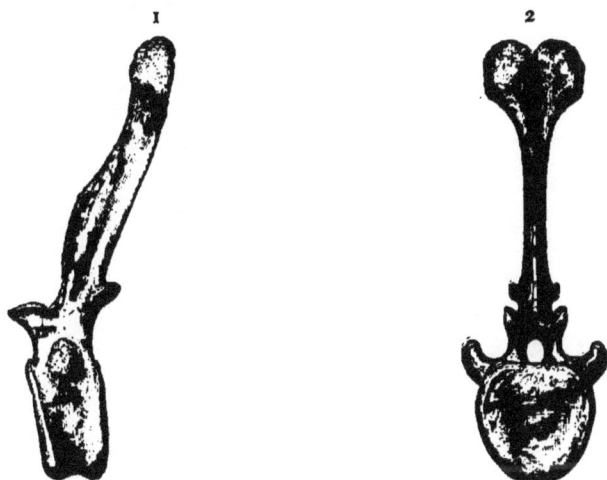

FIG. 26.—Tail vertebræ of *Stegosaurus*. (After Marsh.)
1. Side view. 2. Front view.

tion of the Stegosaur, in which the length of the hind limbs is more apparent, and also they are more free from the body.

Finally, the Stegosaur displays a rather remarkable feature; for a very large chamber was found in the sacrum[1] formed by an enlargement of the spinal cord. The chamber strongly resembled the brain-case in the skull, but was about ten times

[1] The sacrum may be thus defined : the Vertebræ (usually fused together) which unite with the haunch-bones (*ilia*) to form the pelvis.

I

as large !   So this anomalous monster had two sets of brains, one
in its skull, and the other in the region of its haunches ! and
the latter, in directing the movements of the huge hind limbs and

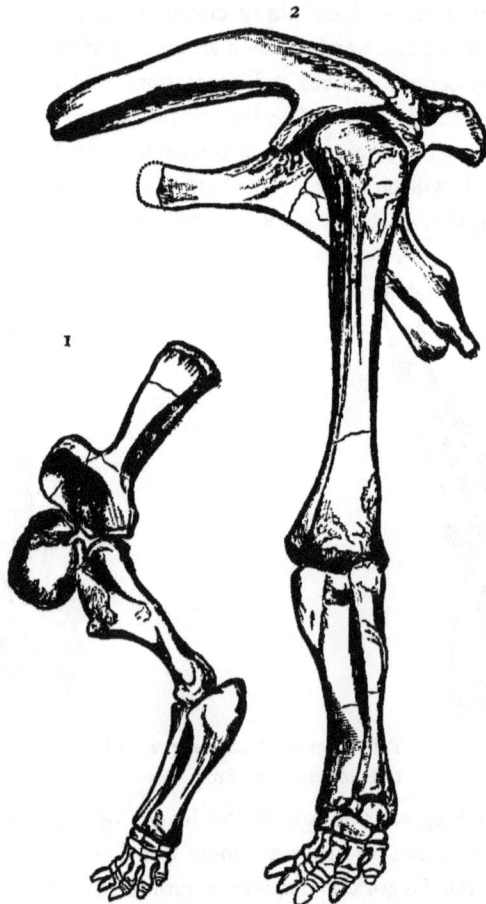

FIG. 27.—Limb-bones of *Stegosaurus.*   (After Marsh.)
1. Fore leg.   2. Hind leg.

tail, did a large share of the work.   The subject is a highly suggestive one, but at present requires further explanation.

On the walls of the fossil reptile gallery at South Kensington the reader will find a large framed drawing of the skeleton of Stegosaurus, kindly sent by Professor Marsh, whose forthcoming monograph will be welcomed by all palæontologists.

The last, and in some ways the strangest of the Dinosaurs, was the Triceratops[1] that flourished in America at the end of the

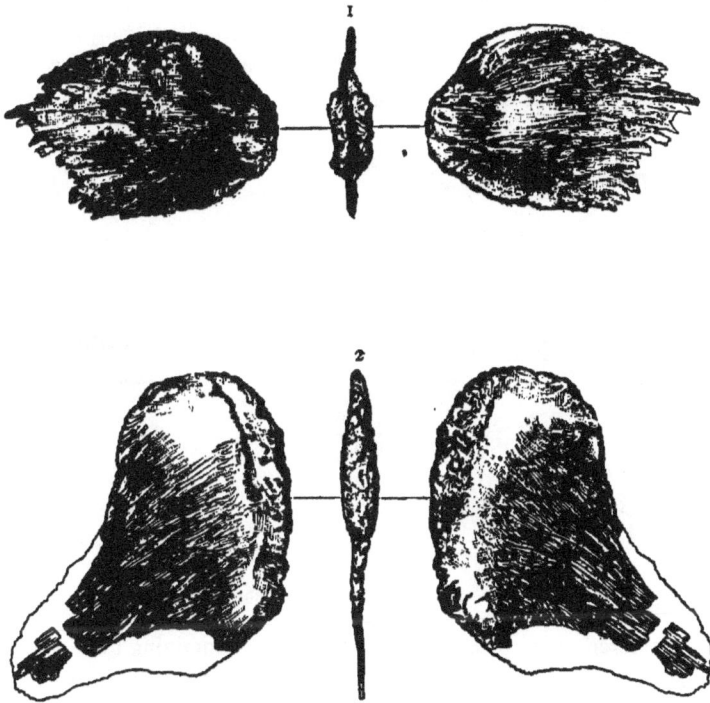

FIG. 28.—1, 2. Plates of Stegosaurus. The middle figures show their thickness. (After Marsh.)

long Mesozoïc era, during the Cretaceous period. The name refers to the three horn-cores found on the skull, which probably supported true horns like those of oxen. Whereas the Stegosaur was provided with quite a small skull, this monster had one of huge

---

[1] Greek—*treis*, three; *ceras*, horn; *ops*, face.

dimensions and remarkable shape (see Figs. 29 and 30).[1]   In the
younger ones it was about six feet long, but in an old individual
must have reached a length of seven or eight feet.   Such a skull
is only surpassed by some whales of the present day.   Twenty
different skulls of this kind have been found, and Professor Marsh
places the horned Dinosaurs in a separate family, to which he has
given the name Ceratopsidæ, or horn-faced.   Their remains come
from the Laramie beds, believed to be of Cretaceous age, but representing a remarkably mixed fauna and flora, so that some have

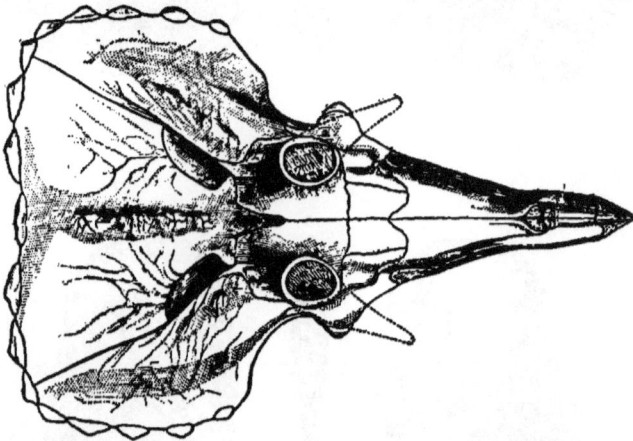

FIG. 29.—Head of *Triceratops*, seen from above.   (After Marsh.)

considered them to be Tertiary.   The strata containing these fossils
are very rich in organic remains, and have yielded not only other
Dinosaurs, but Plesiosaurs, crocodiles, turtles, many small reptiles,
a few birds, fishes, and small mammals.   The Ceratops beds are

[1] This skeleton has not yet been set up in the Yale College Museum,
but will be before long.   Our artist has drawn it as if set up, with a man
standing by for comparison.   In an article in *The Californian Illustrated
Magazine* for April, 1892 (quoted in the *Review of Reviews* for May), an
American writer incorrectly describes this monster as "higher than Jumbo,
and longer than two Jumbos placed in a row."   But the article is altogether
untrustworthy, and the two "restorations" are absurd.

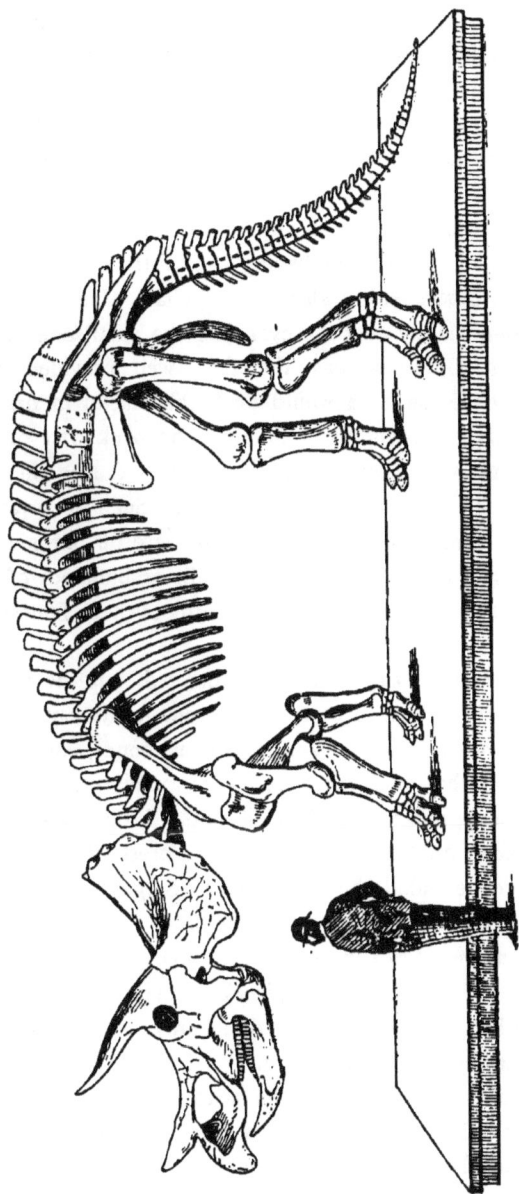

FIG. 30.—Skeleton of *Triceratops prorsus*; length about 25 feet. (After Marsh.)

of fresh-water or brackish origin, and can now be traced for nearly eight hundred miles along the east flank of the Rocky Mountains.

In this Dinosaur we find the fore feet larger than usual in proportion to the hind limbs, and there can be no doubt that it walked on all fours. Its length was about twenty-five feet. All the vertebræ and limb-bones are solid. The brain was smaller in proportion to the skull than in any known vertebrate.

The teeth are remarkable in having two distinct roots. The wedge-like form of the skull is also very peculiar. The two large horns come immediately over the eyes, and the small one above the nose; this Dinosaur was, therefore, well provided with weapons of offence, such as would be highly useful in driving away or wounding carnivorous enemies. The back part of the skull rises up into a kind of huge crest, and this during life was protected by a special fringe of bony plates. Such an arrangement doubtless formed an effective shield to ward off blows when one Triceratops was fighting another, as bulls or buffaloes of the present day fight with their horns. The mouths of these Dinosaurs formed a kind of beak, sheathed in horn.

The body as well as the skull was protected, but the nature and position of the defensive parts in different forms cannot yet be determined with certainty. Various spines, bones, and plates have been found that evidently were meant for the protection of the creature's body, and belonged to the skin. Probably some of these were placed on the back, behind the crest of the skull; some may have defended the throat, as in Stegosaurus. Altogether, Triceratops is very different to any other Dinosaur. One cannot help picturing it rather as a fierce rhinoceros-like animal. In the restoration (Plate XI., Frontispiece) our artist has given it a thick skin, rather like that of the rhinoceros, only indicating small bony plates, etc., here and there.

Professor Marsh thinks that as the head increased in size to bear its armour of bony plates, the neck first, then the fore feet, and then the whole skeleton was specially modified to support it; and he concludes that as these changes took place in the course of the evolution of this wonderful Dinosaur, the head at last became so large and heavy that it must have been too much for the body to bear, and so have led to its destruction! This conclusion, if sound, is a warning against carrying "specialisation" too far. If we wished to write an epitaph on the tomb of the monster, it ought (according to Professor Marsh) to be, "I and my race died of over-specialisation."

After all these various efforts to improve themselves and to perfect their organisation so as to bring it into harmony with

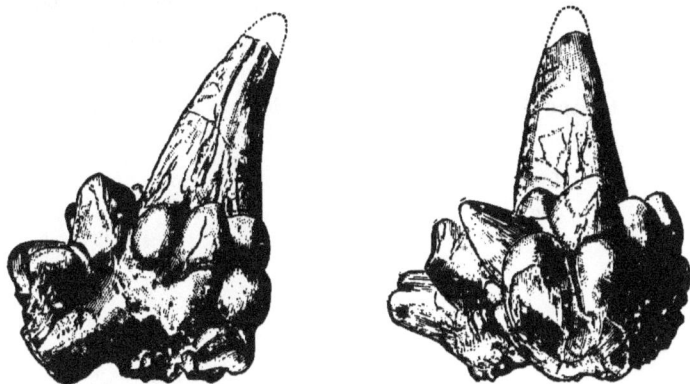

FIG. 31.—Bony spines belonging to the skin of *Triceratops.* (After Marsh.)

their surroundings, or "environment," as the biologists say, it seems rather hard that the Dinosaurs should have been extinguished, and their place in Nature taken by a higher type; but all things have their day, even Dinosaurs.

With regard to the difficulties, hardships, and dangers attending the discovery and transport of the remains, Professor Marsh's concluding remarks may be quoted here, since they give us a glimpse into the nature of his explorations in the far West that have now become so famous. He says, "In conclusion, let me say a word as to how the discoveries here recorded have been accomplished. The main credit for the work justly belongs to my able assistant, Mr. J. B. Hatcher, who has done so much to bring to light the ancient life of the Rocky Mountain regions. I can only claim to have shared a few of the dangers and hardships with him, but without his skill little would have been accomplished. If you will bear in mind that two of the skulls weighed nearly two tons each, when partially freed from their matrix and ready for shipment, in a deep desert cañon, fifty miles from a railway, you will appreciate one of the mechanical difficulties overcome. When I add that some of the most interesting discoveries were made in the hunting-grounds of the hostile Sioux Indians, who regard such explorations with superstitious dread, you will understand another phase of the problem. I might speak of even greater difficulties and dangers, but the results attained repay all past efforts, and I hope at no distant day to have something more of interest to lay before you." [1]

[1] *American Journal of Science*, vol. xli. p. 176.

CHAPTER VIII.

"Geology does better in reclothing dry bones and revealing lost creations than in tracing veins of lead or beds of iron."—RUSKIN.

THE great Ocean of Air was not uninhabited during the long ages of the Mesozoic era, when fishes swarmed in the seas, and reptiles, such as we have attempted to describe in the last five chapters, trod the earth, or swam across lakes and rivers. With such an exuberance of life in various forms, it would indeed have been strange if the atmosphere had only been tenanted by humble little insects like dragon-flies, locusts, or butterflies and moths, all of which we know were living then.

Now, the record of the rocks tells us that one great order of reptiles somehow acquired the power of flying, and flitted about as bats or flying-foxes do now. Since they were undoubtedly reptiles—in spite of certain resemblances to birds—we have ventured to call them "flying dragons," as others have done. The notion of a flying reptile may perhaps seem strange, or even impossible to some persons ; but no one has a right to say such and such a thing "cannot be," or is "contrary to Nature," for the world is full of wonderful things such as we should have considered impossible had we not seen them with our eyes. Charles Kingsley, in his delightful fairy tale, *The Water-Babies*, makes some humorous remarks on that matter, which we may quote here. He says, "Did not learned men too hold, till within the last twenty-five years, that a flying dragon was an impossible monster? And do we not now know that there are hundreds of them found fossil up and down the world? People call them Pterodactyls; but that is only because they are ashamed to call them flying dragons, after denying so long that flying dragons could exist."

The illustrious Cuvier observes that it was not merely in magnitude that reptiles stood pre-eminent in ancient days, but they were distinguished by forms more varied and extraordinary than any that are now known to exist on the face of the earth. Among these extinct beings of ages incalculably remote, are the Pterodactyls,[1] or "wing-fingered" creatures, which had the power of flight, not by a membrane stretched over elongated fingers as in bats, nor by a wing without distinct or complete fingers, as in birds, but by a membrane supported chiefly by a greatly extended little finger, the other fingers being short and armed with claws.

The only reptile now existing which has any power of sustaining itself in the air is the little *Draco Volans*, or "flying lizard," so called; but this can scarcely be regarded as a flying animal. Its hinder pair of ribs, however, are prolonged to such an extent that they support a broad expansion of the skin, so spread out from side to side as to perform the office of a parachute, thus enabling the creature to spring from tree to tree by means of extended leaps; and this it does with wonderful activity.

Many forms of Pterodactyl are known. Some were not larger than a sparrow; others about the size of a woodcock; yet others much larger, the largest of all having a spread of wing (or rather of the flying membranes) of twenty-five feet! It has been concluded that they could perch on trees, hang against perpendicular surfaces, such as the edge of a cliff, stand firmly on the ground, and probably crawl on all fours with wings folded. It may be well at once to point out that the Pterodactyl had no *true* wings like those of a bird, but a thin membrane similar to that of a bat, only differently supported; so it must be understood that, when

---

[1] From the Greek—*pteron*, wing, and *dactylos*, finger.

we use the word " wing," it is not in the scientific sense that we are using it, but in the popular sense, just as we might speak of the wing of a bat, although the bat has no true wing. Figs. 32, 33, 34, and 35 will give the reader some idea of the various forms presented by the skeletons of Pterodactyls, or, as some authorities call them, Pterosaurians (winged lizards). Great differences of opinion have existed among palæontologists as to whether they are more reptilian than bird-like, or even mammalian.

More than a hundred years ago, in 1784, Collini, who was Director of the Elector-Palatine Museum at Mannheim, described a skeleton which he regarded as that of an unknown marine animal. It was a long-billed Pterodactyl from the famous lithographic stone of Solenhofen in Bavaria. The specimen was figured in the *Memoirs of the Palatine Academy.* Collini was able from this specimen to make out the head, neck, small tail, left leg, and two arms; but beyond that, he was at a loss. His conclusion was that the skeleton belonged neither to a bat nor to a bird, and he inquired whether it might not be an amphibian.

In 1809 this specimen came into Cuvier's hands, who at once perceived that it belonged to a reptile that could fly, and it was he who proposed the name Pterodactyl. Until the oracle at Paris was consulted, the greatest uncertainty prevailed, one naturalist regarding it as a bird, another as a bat. Cuvier, with his penetrating eye and patient investigation, combated these theories, supported though they were by weighty authorities. The principal key by means of which he solved the problem, and detected the saurian relationship of the Pterodactyl, seems to have been a certain bone belonging to the skull, known as the quadrate bone. In his great work, *Ossemens Fossiles*, he says, " Behold an animal which, in its osteology, from its teeth to the end of its claws, offers all the characters of the saurians. . . . But it was, at the same time, an animal provided with the means of flight—which, when stationary, could not have made much use of its anterior extremities, even if it did not keep them always folded as birds keep their wings, which nevertheless might use its small anterior fingers to suspend itself from the branches of trees, but when at rest must have been ordinarily on its hind feet, like the birds again ; and also, like them, must have carried its neck sub-erect and curved backwards, so that its enormous head should not interrupt its equilibrium."

Pterodactylus macronyx, or, as it is now called, Dimorphodon macronyx (Fig. 32), was about the size of a raven. It was discovered in 1828 by the late Miss Mary Anning, the well-known collector of fossils from the Liassic rocks that form the cliffs along the coast of Dorsetshire, near Lyme-Regis. This important specimen was figured and described by Dr. Buckland, in the

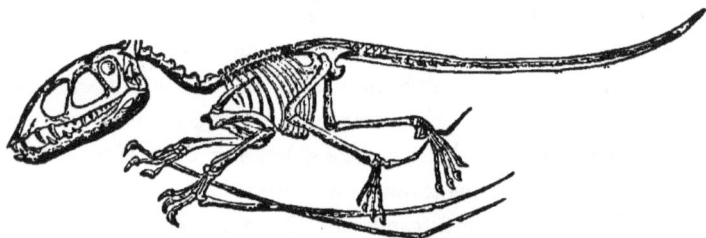

FIG. 32.—Skeleton of *Dimorphodon macronyx.* (After Owen.)

Transactions of the Geological Society. He suggested the specific name macronyx on account of the great length of the claws.

This authority pointed out an unusual provision for giving support and power of movement to the large head at the extremity of a rather long neck, namely, the occurrence of fine long tendons running parallel to the neck-vertebræ. This does not occur in any modern lizards, whose necks are short, and require no such aid to support the head. They are a compensation for weakness that would otherwise arise from the elongation of the neck, supporting, as it did, such a large head. The neck-vertebræ in this species are large and strong, and capable of great flexibility forwards and backwards, so that the creature, by bending its neck during flight into the shape of an S, could throw its head back towards the centre of gravity. The restoration of the skeleton seen in the figure is by Professor Owen. It is probable that this Pterodactyl could walk on the ground with its wings folded, and perhaps it was also capable of perching on trees, by clinging on to their branches with its feet and toes. When the flying membrane was stretched out it must, on account of the long tail to which it was also attached, have presented a triangular shape, somewhat like a boy's kite.

Another genus, also from the lithographic slate of Bavaria, namely, Scaphognathus crassirostris (so called on account of its large beak and jaws), had a very short tail, and its skeleton

FIG. 33.—Skeleton of *Scaphognathus crassirostris.* ⅓ natural size.

looks somewhat clumsy for a creature adapted to fly through the air (Fig. 33).

Pterodactylus spectabilis, from the same strata, also possessed a very short tail, but has a more elegant and bird-like skull. This pretty little flying dragon was only about as large as a sparrow (see Fig. 34). Its neck is comparatively short, with but few joints. The long slender beak was probably sheathed in horn, and the skull in several ways approaches that of a bird. Since there are no teeth in the jaws, we may suppose that it devoured dragon-flies or other insects, such as we know were in existence during the period when the lithographic stone of Bavaria was being deposited. Those forms that were provided with teeth probably devoured such fishes as they could catch by swooping down upon the surface of the water.

Cuvier thought, from the magnitude of their eyes, that Ptero-dactyls were of nocturnal habits. " With flocks of such creatures

FIG. 34.—Skeleton of *Pterodactylus spectabilis.*

flying in the air, and shoals of no less monstrous Ichthyosauri and Plesiosauri swarming in the ocean, and gigantic crocodiles and tortoises crawling on the shores of the primæval lakes and rivers —air, sea, and land must have been strangely tenanted in these early periods of our infant world." [1]

[1] Buckland, *Bridgewater Treatise.*

It was thought at one time that Birds differed from Pterodactyls in the absence of teeth ; but this only holds good for modern birds. If we go back to the Mesozoic age, we find that birds at that time did possess teeth. The oldest known bird, the Archæopteryx, had teeth in its jaws, and presents some very striking points of resemblance to reptiles. But if we compare the skeleton of a Pterodactyl (such as the P. spectabilis, now under consideration) with that of a bird, we shall see in its fore limbs certain very obvious differences. A bird never has more than three fingers in its hand or wing (viz. the thumb and next two digits), and the bones that support these fingers, corresponding to the bones in the palm of a human hand, are joined together. Neither of the bones corresponding to our fingers are much elongated, and of these the longest is that which corresponds to the thumb. But, on referring to the skeleton of our Pterodactyl, we find that it has four fingers, three of which are fairly developed and furnished with claws, while the outermost one is enormously elongated. This is believed to correspond to the little finger of the human hand, while the thumb seems to be represented by a small bone seen at the wrist. It was this long outside finger that chiefly served to support the flying membrane of the Ptero- dactyl. For this and other reasons, we are forbidden to look upon these creatures as relatives of birds. Again, all birds that can fly possess a " merrythought," or furculum ; and such is not found in the Pterodactyl.

As we have already remarked, some authorities, when these creatures were first brought to light, considered them to be mammals, as bats are. But equally conclusive arguments may be brought forward against that view. All mammals have the skull jointed to the backbone by two articulations, known as " condyles," whereas Pterodactyls have only one—in that respect resembling reptiles and birds.

Also there are important differences in the structure of their jaws, showing that they are constructed on the reptilian plan, and not on that of the mammal.

In order to give rapid movement to their wings during flight, they had powerful muscles in the region of the chest. These were attached to a shield-like breast-bone provided with a keel—as in birds. But this bird-like feature is only a necessary provision to enable them to fly, and does not point to any relationship.

In the year 1873 was discovered, in the lithographic stone of Bavaria, at Eichstädt, a very beautiful new form of Pterodactyl. This was the Rhamphorhynchus phyllurus. The specimen is in a remarkable state of preservation; for the bones of the skeleton are nearly all in position, while those of both wings show very perfect impressions of the membranes attached to them. Its long tail supported another small leaf-like membrane, which was evidently used as a rudder in flight (see Fig. 35). The dis-

FIG. 35.—Skeleton of *Rhamphorhynchus phyllurus*, with delicate impressions of the flying membranes. (After Marsh.)

covery of this valuable specimen attracted much attention at the time. It was bought, by telegram, for Professor Marsh, and so secured for the Yale College Museum; but a cast may be seen at South Kensington (Wall-case, No. 1, Gallery IV. on plan).

Any one who looks carefully at the beautiful impressions of the wings of this specimen can see that they must have been produced by a thin smooth membrane, very similar to that of bats. When this elegant little creature was covered up by the fine soft mud that now forms the lithographic stone, its wings were partly folded, so that the membranes were more or less contracted into folds, like an umbrella only partly open. These appear to have been attached all along the arm and to the end of the long finger. They then made a graceful curve backward to the hind foot, and probably were continued beyond the latter so as to join the tail. With its graceful pointed wings and long tail, this little flying saurian must have been a beautiful object, as it slowly mounted upwards from some cliff overlooking the Jurassic seas. (See Plate XII.)

Like those already described, it was provided with four short-clawed fingers, as well as the one which mainly supported its wing. Some of the Continental museums contain good collections of fossil Pterodactyls; but the largest collection in the world is that of Yale College, where Professor Marsh declares there are the remains of six hundred individuals from the American Cretaceous rocks alone !

Some of the fragmentary remains from our Cambridge Green-sand formation indicate Pterodactyls of enormous size. Thus

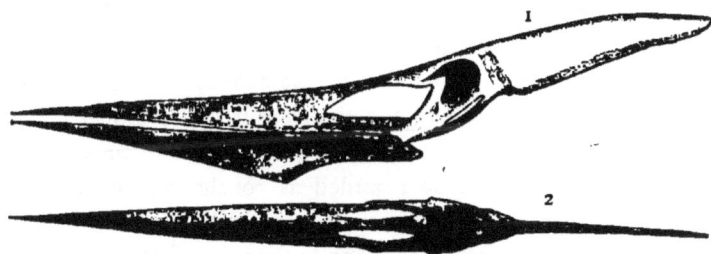

FIG. 36.—Skull of *Pteranodon*. 1. Side view. 2. Top view. (After Marsh.)

the neck-vertebræ of one species measure two inches in length, while portions of arm-bones are three inches broad. It is probable that the creatures to which these bones once belonged measured eighteen or twenty feet from tip to tip of the wings. Other also fragmentary remains from the chalk of Kent testify to

K

the existence of Pterodactyls during that period fully equal in size.

But the largest Pterodactyls hail, like so many other big things, from America. Professor Marsh tells us of monsters in his famous collection with a spread of wings of twenty to twenty-five feet! These large forms had no teeth in their jaws, and their skulls are of a peculiar form. The long-pointed jaws were probably sheathed in horn during life, as in birds (see Fig. 36). According to Marsh, these toothless forms (which he calls Pteranodonts) were mostly of gigantic size. With regard to their food it is almost vain to speculate; but if they *did* prey upon fishes, they must have had a capacious mouth and gullet, and must have swallowed their prey whole, after the fashion of pelicans. But we doubt if they had the peculiar pouch possessed by those birds. In the absence of more complete accounts of the large forms the artist has only attempted to restore the small ones. (See Plate XII., showing four different kinds.)

Whether Pterodactyls were cold-blooded or warm-blooded is a question on which the authorities are not agreed. Professor Owen argued from the absence of feathers that they could not have been warm-blooded. But, in spite of this great authority, who has defended his opinion somewhat strongly, there are others who argue that the amount of work involved in sustaining a Pterodactyl in the air make it highly probable that it was warm-blooded. The absence of feathers to retain the heat of the body need not be regarded as conclusive, for bats are warm-blooded animals, and in their case the heat of the body is retained by a slight downy covering to the skin. Such a covering may have protected the bodies of Pterodactyls, and we could not expect to see any trace of it in the Bavarian specimen of Rhamphorhynchus referred to above. An important fact bearing on this question is the discovery of perforations in the bones of these animals very similar to those seen in birds. Now, birds have a wonderful system of respiration, or breathing. The air

they breathe passes, not into their lungs only, but penetrates to the remotest parts of their system, filling their very bones with life, and endowing them with activity and animation adapted to their active aërial existence. It may, therefore, be argued that Pterodactyls breathed much in the same way; that their bones, too, were supplied with air by an elaborate system of air-sacs, and that they had lungs like those of birds. We cannot, however, stop there, but are led on by physiological reasoning to conclude that the circulation of the blood must have been rapid, and that the heart was like that of birds and mammals, four-celled. It would therefore follow—since birds and mammals are warm-blooded—that Pterodactyls were also. Such, at least, is the view of Professor H. G. Seeley, who says of the Cambridge specimens, "That they lived exclusively upon land and in air is improbable, considering the circumstances under which their remains are found. It is likely that they haunted the sea-shores, and, while sometimes rowing themselves over the water with their powerful wings, used the wing-membranes, as the bat does, to enclose their prey, and bring it to the mouth.

"The large Cambridge Pterodactyls probably pursued a more substantial prey than dragon-flies. Their teeth are well suited for fish, but probably fowl and small mammals, and even fruits, made a variety in their food. As lord of the cliff, it may be presumed to have taken toll of all animals that could be conquered with tooth and nail. From its brain it might be regarded as an intelligent animal. The jaws present indications of having been sheathed with a horny covering."

Probably the large Pterodactyls of the Cretaceous period, soaring like albatrosses and giant petrels over the surface of the ocean, co-operated with the marine reptiles, such as Ichthyosaurs, Plesiosaurs, crocodiles, and others, as those sea-birds now do with the whales, porpoises, and dolphins, in reducing the excessive numbers of the teeming tribes of fishes, and in maintaining the balance of oceanic life.

With regard to the place of Pterodactyls in the animal kingdom, Professor Seeley places them as a distinct sub-class, side by side with birds, and between mammals and reptiles, thus—

Mammalia.

Ornithosauria   Aves.

Reptilia.

The name Ornithosauria (bird-lizards) is frequently used instead of the other name, because it expresses the idea of their being partly saurian, and partly bird-like.

They flourished from the period of the Lias to that of the Chalk; and then, like so many other strange forms, seem to have suddenly disappeared.

CHAPTER IX.

SEA-SERPENTS.

"Sand-strewn caverns, cool and deep,
  Where the winds are all asleep;
  Where the spent lights quiver and gleam,
  Where the salt weed sways in the stream;
  Where the sea-beasts, ranged all round,
  Feed in the ooze of their pasture-ground;
  Where the sea-snakes coil and twine,
  Dry their mail, and bask in the brine."
                The Forsaken Merman.

IT has been said that everything on earth has its double in the water. Are there not water-beetles, water-scorpions, water-rats, water-snakes, sea-lions, sea-horses, and a host of other living things, whether plants or animals, bearing some sort of resemblance to others that live on land? Then why not sea-serpents? The great controversy of the sea-serpent, that has so often been discussed in the newspapers, need not be considered here. We are dealing not with the present, but with the past; and whether or no the wonderful sailors' yarns of sea-serpents can be regarded as authentic, even in a single case, we can offer our readers infallible proof that, during the so-called "Age of Reptiles," certain monstrous saurian animals flourished in considerable abundance, which, though not true serpents, nevertheless must have borne a striking resemblance to such, as they cleaved he waters of primæval seas.[1]

[1] See an interesting little work, entitled, *Sea-Monsters Unmasked*, by H. Lee (Clowes and Sons). Appendix II. contains some extracts therefrom.

The modern evolutionist believes that snakes are descended from lizards, possessing, as usual, four legs ; that some primitive form of lizard with very small legs appeared on the scene, and found that it could better move along by wriggling its body and pushing with its ribs than by walking. So, in course of time, a race of lizards without legs arose ; these, by Natural Selection, and perhaps other means, became more and more elongated, so that they could move faster than their ancestors, and glide out of harm's way more effectually. Thus was the snake evolved from a lizard.

Now, in the great geological museum of the stratified rocks, there have been discovered skeletons of marine reptiles, which propelled themselves chiefly by means of their tails and elongated bodies, rather than by their limbs. The limbs were not discarded entirely as in the case of the serpents, but were useful in their way as the fins of fishes are. Perhaps, therefore, we may be justified in calling these ancient monsters sea-serpents, in consideration of their long thin bodies ; for they certainly would be called by that name if now living.

Strictly speaking, they were not serpents, but more or less like some of the extinct saurians described in chap. iv. The name, however, has been adopted by geologists, and is useful in so far as it serves to remind us of their very peculiar shape and structure. Remains of these strange creatures have been found both in Europe and America.

One of the earliest discoveries of remains of a fossil sea-serpent was made by M. Hoffman, a Dutch military surgeon, in the year 1770. Maestricht, a city in the interior of the Netherlands, situated in the valley of the Meuse, stands on certain strata of limestone and sandstone, belonging to the Upper Chalk. Extensive quarries have, for many centuries, been worked in the sandstone, especially in the eminence called St. Peter's Mount, which is a cape or headland between the Meuse and the Jaar. This elevated plateau extends for some distance towards

Liége, and presents an almost perpendicular cliff towards the Meuse. From the extensive works that have so long been carried on, immense quantities of stone have been removed, and the centre of the mountain is traversed by galleries, and hollowed by vast excavations. Innumerable fossils, such as marine shells, corals, crustaceans, bones and teeth of fishes, have been obtained from this rock. But St. Peter's Mount is now chiefly cele-brated for the discovery of the bones and teeth of a huge saurian, to which Mr. Conybeare has given the name Mosasaurus, on account of its connection with the river Meuse. M. Hoffman had long been an assiduous collector of fossils from this neigh-bourhood, and he had the good fortune to obtain the famous specimen on which this genus is founded.

It was at first considered, by M. Faujas St. Fond, to be a crocodile; but Cuvier and Camper formed a different and better conclusion. Perhaps no fossil ever had such a remarkable history as this one, as the following account, from M. Faujas St. Fond's work on the fossils of St. Peter's Mount,[1] will show.

"Some workmen, on blasting the rock in one of the caverns of the interior of the mountain, perceived, to their astonishment, the jaws of a large animal attached to the roof of the chasm. The discovery was immediately made known to M. Hoffman, who repaired to the spot, and for weeks presided over the arduous task of separating the mass of stone containing these remains from the surrounding rock. His labours were rewarded by the successful extrication of the specimen, which he conveyed in triumph to his house. This extraordinary discovery, however, soon became the subject of general conversation, and excited so much interest, that the canon of the cathedral which stands on the mountain resolved to claim the fossil, in right of being lord of the manor; and succeeded, after a long and harassing lawsuit, in obtaining this precious relic. It remained for years in his

---

[1] *Histoire Naturelle de la Montagne de St. Pierre.* This account is given by Dr. Mantell, in his *Petrifactions and their Teaching*, 1851.

possession, and Hoffman died without regaining his treasure, or receiving any compensation.   At length the French Revolution broke out, and the armies of the Republic advanced to the gates of Maestricht.   The town was bombarded ; but, at the suggestion of the committee of savans who accompanied the French troops to select their share of the plunder, the artillery was not suffered to play on that part of the city in which the celebrated fossil was known to be preserved.   In the mean time, the Canon of St. Peter's, shrewdly suspecting the reason why such peculiar favour was shown to his residence, removed the specimen, and con-cealed it in a vault ; but when the city was taken, the French authorities compelled him to give up his ill-gotten prize, which was immediately transmitted to the Jardin des Plantes, at Paris, where it still forms one of the most striking objects in that magnificent collection."

Dr. Mantell quotes the Frenchman's remark on this transaction :
" *La Justice, quoique tardive, arrive enfin avec le temps :* " but adds, " The reader will probably think that, although the reverend canon was justly despoiled of his ill-gotten treasure, the French commissioners were but very equivocal representatives of *Justice!* "

The beautiful cast (Fig. 37) at South Kensington (Fossil Reptile Gallery, Wall-case 8) was presented to Dr. Mantell by Baron Cuvier in 1825.   It consists of both jaws, with numerous teeth, and some other parts (see Fig. 38).   The length is about four and a half feet.   This nearly perfect head was for a time a stumbling-block to many naturalists, some of whom were of opinion that it belonged to a whale.   Cuvier and others considered it to be a kind of link between the Iguanas and the Monitors.[1]

The entire backbone of the Maestricht animal appears to have

[1] The Monitors are a family of large lizards inhabiting the warmer parts of Africa and Asia.   They live near the banks of rivers, and some are altogether aquatic.   They often devour the eggs of crocodiles and aquatic birds.   The Nile Monitor, or Varanus, grows to a length of six feet.

consisted of one hundred and thirty-one vertebræ, of which ninety-seven belonged to the tail. The total length of the skeleton is

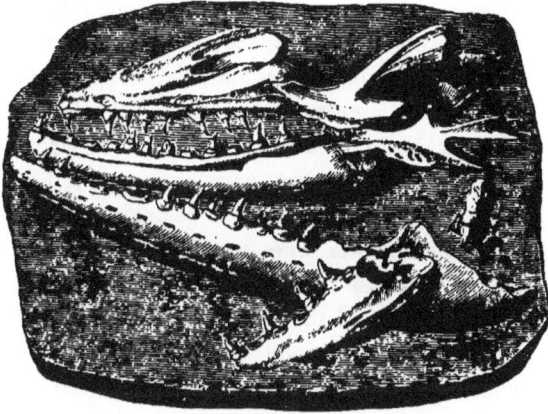

FIG. 37.—Skull of *Mosasaurus Hoffmanni.* The original is 4½ ft. by 2¼ ft.

estimated at twenty-four feet, and the head was about one-sixth of the total length. The tail is only ten feet long, whereas in a

FIG. 38.—Teeth of Mosasaurus (half natural size). 1ᵃ, 2ᵃ, transverse sections of the teeth.

crocodile the tail exceeds the length of the body. Although in his day the limbs of the Mosasaurus were imperfectly known, Cuvier rightly considered them to be adapted for swimming, and,

with his usual foresight, concluded that this monster was a marine
reptile of great strength and activity, having a large tail flattened
vertically and capable of being moved from side to side with
such force and rapidity as to be a powerful organ of propulsion,
capable of stemming the most agitated waters.   The large conical
recurved teeth, the largest of which was nearly three inches long,
are well seen in Figs. 37 and 38. Dr. Mantell was fortunate enough
to find, in the year 1820, some vertebræ from the English Chalk
near Lewes, which were identified as belonging to a Mosasaurus.

In 1831 a portion of a lower jaw with large conical teeth was
discovered in the Chalk near Norwich.   But these teeth were not
quite similar to those of the Maestricht specimen, and Professor
Owen therefore founded upon them the new genus Leiodon.[1]
But Leiodon must have been very similar to Mosasaurus.

FIG. 39.—Lower tooth of *Leiodon*.   1. Side view.   2. Profile.

Of late years many fine specimens have been discovered in
North America, and the labours of Leidy, Marsh, and Cope have

[1] Greek—*leios*, smooth, and *odous*, tooth.

been of the greatest service in completing our knowledge of this strange group of saurians. In the American Cretaceous seas they ruled supreme, as their numbers, size, and carnivorous habits enabled them easily to vanquish all rivals. Probably some of them were seventy-five feet in length, the smallest being ten or twelve feet long. In the inland Cretaceous sea from which the Rocky Mountains were beginning to emerge, these ancient sea-serpents abounded; and many were entombed in its muddy deposits. On one occasion, as Professor Marsh rode through a valley washed out of this old ocean bed, he observed no less than seven different skeletons of these monsters in sight at once! The same authority mentions that the Museum of Yale College contains remains of not less than 1400 distinct individuals. In some of these the skeleton is nearly if not quite complete; so that every part of its structure can be determined with almost absolute certainty.

According to Professor Cope of Pennsylvania University, who has made a special study of this group of extinct saurians, fifty-one species have been discovered in North America, in the States of New Jersey, Alabama, Kansas, North Carolina, Mississippi, and Nebraska. The same authority has shown that they were characterised by a wonderful elongation of form, especially of the tail; that their heads were large, flat, and conical in shape, with eyes directed partly upward; that they were furnished with two pairs of paddles like the flippers of a whale. With these flippers, and the eel-like strokes of their flattened tail, they swam with considerable speed. Like snakes, they were furnished with four rows of formidable teeth on the roof of the mouth, which served admirably for seizing their prey.

But the most remarkable feature in these creatures was the arrangement for permitting them to swallow their prey whole, in the manner of snakes. Thus each half of the lower jaw was articulated at a point nearly midway between the ear and the chin, so as to greatly widen the space between the jaws, and

Professor Cope thinks that the throat must consequently have been loose and baggy.

Professor Cope, however, in giving the name Pythonomorpha to this ancient group, has pressed his views too far, and dwelt unduly on their supposed relationship with serpents. Other authorities regard them as essentially swimming lizards, with four well-developed paddles; and this is probably the right view to take of them.

The following graphic account of the region where Professor Cope has discovered the skeletons of many sea-serpents, and of their habits and aspect when alive, is taken from his well-known work on the Cretaceous Vertebrata of the West.[1] After describing this region as a vast level tract between the Missouri and the Rocky Mountains, he says, "If the explorer searches the bottoms of the rain-washes and ravines, he will doubtless come upon the fragment of a tooth or jaw, and will generally find a line of such pieces leading to an elevated position on the bank or bluff, where lies the skeleton of some monster of the ancient sea. He may find the vertebral column running far into the limestone that locks him in his last prison; or a paddle extended on the slope, as though entreating aid; or a pair of jaws lined with horrid teeth, which grin despair on enemies they are helpless to resist; or he may find a conic mound, on whose apex glisten in the sun the bleached bones of one whose last office has been to preserve from destruction the friendly soil on which he reposed. Sometimes a pile of huge remains will be discovered, which the dissolution of the rock has deposited on the lower level; the force of rain and wash having been insufficient to carry them away."

But the reader inquires, "What is the nature of these creatures thus left stranded a thousand miles from either ocean? How came they in the limestone of Kansas, and were they

[1] *Report of the United States Geological and Geographical Survey of the Territories*, vol. ii., 1875 (*Cretaceous Vertebrata*).

denizens of land?" These creatures lived in the Cretaceous period. The remains found in this region were mostly those of reptiles and fishes. Thirty-five species of reptiles are known from Kansas alone, representing six orders, and varying in length from ten to eighty feet. One was terrestrial, four were fliers, the rest inhabited the ocean. "When they swam over what are now the plains, the coast-line extended from Arkansas to near Fort Riley, on the Kansas River, and, passing a little eastward, traversed Minnesota to the British possessions, near the head of Lake Superior. The extent of sea to the westward was vast, and geology has not yet laid down its boundary; it was probably a shore now submerged beneath the waters of the North Pacific."

Other very elongated marine reptiles of this period, but with much thicker bodies, are called, by Professor Cope, Elasmosaurs. In this group, which is not yet fully worked out, occur such genera as Cimoliosaurus, Polycotylus, Polyptychodon, and others. But it seems a pity that they should be in any way separated from the Plesiosaurs, which they strongly resemble (see chap. iv., Plate III.). Though not sea-serpents, we have introduced them here because they flourished at the same time, and lived in the same seas with the Mosasaurs and other forms of that group. The very large teeth, with strongly marked ridges, of the Polyptychodon are abundant in the Cambridge Greensand that underlies the chalk, and represent a very huge animal.

In our illustration, Plate XIII., the artist has represented the Elasmosaurus[1] (of Cope) with its long thin neck stretched out in search of food on the bed of the sea. Professor Cope—thus describing this monster, in language which seems somewhat fanciful—says, "Far out on the expanse of this ancient sea might have been seen a huge snake-like form, which rose above the surface, and stood erect, with tapering throat and arrow-shaped head, or swayed about, describing a circle of

[1] Greek—*elasmos*, plate; *sauros*, lizard : probably on account of the shape of the paddles.

twenty feet radius above the water. Then plunging into the depths, naught would be visible but the foam caused by the disappearing mass of life. Should several have appeared together, we can easily imagine tall, flexible forms rising to the height of the masts of a fishing-fleet, or like snakes twisting and knotting themselves together. This extraordinary neck—for such it was—rose from a body of elephantine proportions. The limbs were probably two pairs of paddles, like those of Plesiosaurus, from which this diver chiefly differed in the arrangement of the bones of the breast. In the best-known species twenty-two feet represent the neck in a total length of fifty feet. This is Elasmosaurus platyurus (Cope), a carnivorous sea-reptile, no doubt adapted for deeper waters than many of the others. Like the snake-bird of Florida, it probably often swam many feet below the surface, raising the head to the distant air for breath, then withdrawing it, and exploring the depths forty feet below, without altering the position of its body. From the localities in which the bones have been found in Kansas, it must have wandered far from land; and that many kinds of fishes formed its food is shown by the teeth and scales found in the position of its stomach."

But to return to the sea-serpents. Mosasaurus is now known to have been a long slender reptile, with a pair of powerful paddles in front, a moderately long neck, and flat pointed head. The tail was very long—flat and deep—like that of a great eel. Mosasaurus princeps is computed to have been seventy-five to eighty feet long. Clidastes was another genus of long and slender shape, one species of which reached a length of forty feet. Some forms of sea-serpent had sclerotic plates in the eye, such as we found in the fish-lizard, or Ichthyosaurus (p. 46), but the announcement that their bodies were protected by bony plates has turned out to be a mistake, and the supposed plates really belonged to the eye.

Leiodon proriger (Cope) was abundant in the old North

American Cretaceous sea, and reached a length of seventy-five feet. It had a long projecting muzzle, somewhat like the snout of a sturgeon. Platecarpus and Tylosaurus had peculiarly sharp-pointed heads (see Fig. 40).

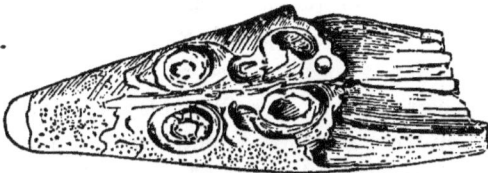

FIG. 40.—Snout of Tylosaurus. (After Marsh.)

A few words may be added here with regard to Professor Cope's important discovery of Leiodon—a genus already alluded to as having been founded by Sir Richard Owen. The type specimen of Leiodon dyseplor,[1] which first indicated the characters of this wonderful species, was obtained from the yellow beds of the Niobrara epoch of the Jornada del Muerto, near Fort McRae, New Mexico. The greater part of the remains have been described by Professor Leidy. But a second specimen, more complete in all respects, was discovered by Professor Cope's exploring party during an expedition from Fort Wallace, Kansas, in 1871. This specimen he has fully described and figured in the report already referred to (p. 140). It is a very instructive specimen, including fifty of the vertebræ from all parts of the vertebral column, a large part of the cranium, with teeth, as well as important limb-bones. These precious relics were excavated from a chalk " bluff," or high bank. Fragments of the jaws were seen lying on the slope, and other portions entered the shale. On being followed, a part of the skull was taken from beneath the roots of a bush, and the vertebræ and limb-bones were found farther in. The series of vertebræ, after extending some way along the face of the bluff, finally turned into the hill, and were

[1] We retain the old spelling with the *e* as being nearer to the Greek, although Professor Cope writes it " Liodon."

followed as far as time would permit, but part of the tail series had to be left. In size, the vertebræ of this enormous sea-serpent exceed those of Mosasaurus brumbyi. The latter has hitherto been the largest known species of the order of Pythonomorphs, exceeding twofold in its measurements the M. giganteus of Belgium. So the present reptile is much larger in its dimensions than the New Jersey species called maximus by Professor Cope. "If, as appears certain," says the professor, "the Mosasauroid discovered by Webb measures seventy-five feet in length, and the M. maximus measured eighty, the Leiodon dyspelor must have been the longest reptile known, and approaches very nearly the extreme of the mammalian growth seen in the whales, though, of course, without their bulk. Such monsters may well excite our surprise, as well as our curiosity, in the inquiry as to their source of food-supply, and what the character of those contemporary animals preserved in the same geologic horizon."

In our illustration, Plate XIII., the artist has endeavoured to realise the outward aspect of the two genera of sea-serpents, Mosasaurus and Clidastes. The fishes which they are pursuing are well-known genera from the English Chalk, such as Beryx.

Ten species of Clidastes have been unearthed from the Kansas strata. They did not reach such a size as the Leiodons, but were of elegant and flexible build, the largest species, C. cineriarum, reaching a length of forty feet (see Fig. 41). A smaller species, of elegant proportions, has been called C. tortor (Cope). Its slenderness of body was remarkable, and the large head was long and lance-shaped. Its lithe movements doubtless helped it to secure many fishes. It was found coiled up beneath a ledge of rock, with its skull lying undisturbed in the centre.

The accounts given by Professor Cope of his explorations and the difficulties encountered in procuring the valuable specimens on which his conclusions are based, are most interesting, and such as every fossil-hunter will appreciate. We, in England, who visit clay pits, stone quarries, railway cuttings, etc., during a morning or an afternoon walk, and return home at our leisure with a few small specimens in our pockets, or in a bag at our back, can hardly realise how arduous must be the work of finding, digging out, and transporting for such long distances the remains of the monsters of Kansas and other parts of North America.

The following extracts have been selected from Professor Cope's report, with a view to illustrating the nature of the explorations undertaken. "The circumstances attending the discovery of one of these will always be a pleasant recollection to the writer. A part of the face, with teeth, was observed projecting from the side of a bluff by a companion in exploration, Lieutenant James H. Whitten, United States Army, and we at once proceeded to follow up the indication with knives and picks. Soon the lower jaws were uncovered, with their glistening teeth, and then the vertebræ and ribs. Our delight was at its height when the bones of the pelvis and part of the hind limb were laid bare, for they had never been seen before in the species, and scarcely in the order. While lying on the bottom of the Cretaceous sea, the carcase had been dragged hither and thither by the sharks and other rapacious animals, and the parts of the skeleton were displaced and gathered into a small area. The massive tail stretched away into the bluff, and, after much laborious excavation, we left a portion of it to more persevering explorers."

"The discovery of a related species, Platecarpus coryphæus (Cope), was made by the writer under circumstances of difficulty peculiar to the plains.

FIG. 41.—Skeleton of *Clidastes cineriarum*; length 40 feet. (After Cope.)

After examining the bluffs for half a day without result, a few bone fragments were found in a wash above their base. Others led the way to a ledge forty or fifty feet from both summit and foot, where, stretched along in the yellow chalk, lay the projecting portions of the whole monster. A considerable number of vertebræ were found preserved by the protective embrace of the roots of a small bush, and, when they were secured, the pick and knife were brought into requisition to remove the remainder. About this time, one of the gales, so common in that region, sprang up, and striking the bluff fairly, reflected itself upwards. So soon as the pick pulverised the rock, the limestone dust was carried into eyes, nose, and every available

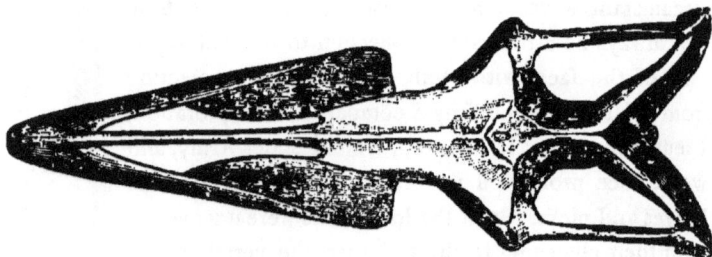

FIG. 41a.—Skull of *Platecarpus.* Upper Cretaceous. North America.
(After Cope.)

opening in the clothing. I was speedily blinded, and my aid disappeared in the cañon, and was seen no more while the work lasted. A handkerchief tied over the face, and pierced by minute holes opposite the eyes, kept me from total blindness, though dirt in abundance penetrated the mask. But a fine relic of Creative Genius was extracted from its ancient bed, and one that leads its genus in size, and explains its structure."

"On another occasion, riding along a spur of yellow chalk bluff, some vertebræ lying at its foot met my eye. An examination showed that the series entered the rock, and, on passing round to the opposite side, the jaws and muzzle were seen projecting from it, as though laid bare for the convenience of the geologist. The spur was small and of soft material, and we speedily removed it in blocks, to the level of the reptile, and took out the remains as they lay across the base from side to side."

In taking leave of the "Age of Reptiles," we cannot but marvel greatly at the diversity of forms assumed by the various orders of this class, their strange uncouth appearance, their assumption, in some cases, of characters only known at the present day among the mammals, their great abundance, and the perfect state in which their remains have been preserved in the stratified rocks of various parts of the world. And the reader may naturally ask, "How is it that so many types have disappeared altogether, leaving us out of a total of at least nine orders, only four, viz. those represented by crocodiles, lizards, snakes, and turtles?" To such a question we can only answer that the causes of the extinction of plants and animals in the past are not yet known. Climate, geographical conditions, food-supply, competition, with other causes, doubtless operated then as now; but if there is one clear lesson taught by the record of the rocks, it is this—that there has been at work from the earliest periods a Law of Progress, so that higher types, coming in at certain stages, have ousted the lower types, sometimes only partially, sometimes completely. But why the Dinosaurs, for instance, perished entirely, while the crocodiles survived to the present day, no one can yet explain. We can see no reason, however, why such problems as these should not be solved in the future by the co-operating labours of naturalists and geologists.

In the great onward and upward struggle for existence, higher types have supplanted lower ones; and, in accordance with this biological truth, we find that in the next era (known as the Tertiary or Cainozoic) the mammal held the field while the reptile took a subordinate place.

[handwritten annotation]

CHAPTER X.

SOME AMERICAN MONSTERS.

"Geology, in the magnitude and sublimity of the objects of which it treats, ranks next to Astronomy in the scale of the Sciences."—SIR JOHN F. W. HERSCHEL.

WITH the advent of the Cainozoic or Tertiary era, we enter upon the "Age of Mammals," when great quadrupeds suddenly came upon the scene. The place of the reptile was now taken by the mammal. In the long previous era this higher type of life was not altogether wanting, but as far as the geological record is yet known, it appears only to have been represented by a few primitive little creatures, probably Marsupials, whose jaw-bones have been discovered in the New Red Sandstone, and the Stonesfield Oolite.[1]

Geology tells of a great gap between the highest rocks of the Cretaceous period and the lowest group of the succeeding Eocene period (see Table of Strata, Appendix I.). This gap, or break, testifies to a very long interval of time during which important geographical and other changes took place; and consequently we find in Eocene rocks (at the base of the Cainozoic series) a very different fauna and flora to that which is preserved in the Chalk formation.

The researches of Cuvier among the fossils collected from Eocene rocks in the neighbourhood of Paris, especially the

[1] The English Cretaceous rocks, previously thought to be destitute of mammalian remains, have quite recently yielded teeth belonging to some small mammal. These were found in Wealden strata.

Gypseous series of Montmartre, revealed the existence of a very extensive fauna, especially of new types of mammals ; and his restoration of the Palæotherium, a tapir-like animal, and other forms, created a vast amount of interest, and greatly stimulated the study of extinct animals. As we have already remarked, the science of palæontology may be said to have been founded by Cuvier (see Introduction, p. 5).

But now the scene shifts once more from Europe to the wilds of the Far West. American geologists tell us that a long time ago (during the Eocene period) there was a great tropical lake in the Wyoming territory, on the borders of which roamed, amidst luxuriant vegetation, a large number of strange and primitive quadrupeds, together with many other forms of life. The most wonderful group of animals that haunted the shores of this lake, or series of lakes, was the Dinocerata so fully described by Professor Marsh, in his exhaustive monograph.[1] The name implies that they were terrible horned monsters, but whether Nature provided them with true horns, like those of horned cattle to-day, is at least open to doubt.

Fig. 42 shows the skeleton of one of these, namely, Tinoceras ingens. Its length was about 12 feet without the tail. Its weight, when alive, is calculated to have been six thousand pounds, or about two tons and three quarters.

Plate XIV. is a restoration of the Tinoceras, made by our artist, after much consideration and careful study of the valuable cast exhibited in the Natural History Museum at South Kensington, which was generously presented by Professor Marsh (Gallery I. Case MM on plan). In planning this and other restorations, both artist and author have received valuable assistance from Dr. Henry Woodward, F.R.S., Keeper of the Geological Department of the Museum, who is ever ready to help with his great knowledge those who come to consult him.

[1] *The Dinocerata*, a monograph by O. C. Marsh, *United States Geological Survey*, vol. x.

There may be differences of opinion among palæontologists as to the appearance presented by this formidable creature when alive, and no doubt the nature of the skin must always be more or less a matter of conjecture in such cases, but we venture to hope that the restoration here given, based as it is upon Mr. Smit's thorough acquaintance with living animals and Professor Marsh's description, will meet with a favourable verdict.

Looking at the skeleton, one is struck with a certain resemblance to the rhinoceros on one hand, and to the elephant on the other. The legs are very elephantine, and the feet must have

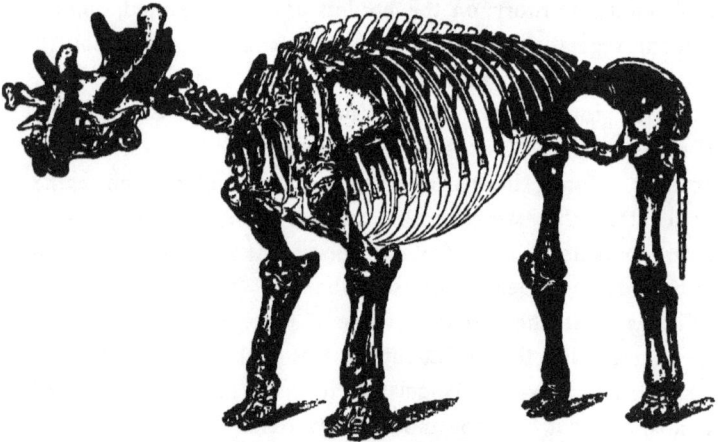

Fig. 42.—Skeleton of *Tinoceras ingens.* (After Marsh.)

been covered with thick pads, but the body reminds one more of the rhinoceros; and yet, again, there is some suggestion of the hippopotamus. The eye was small and deep set, as in the rhinoceros. In the upper jaw the two canine teeth are developed into dagger-shaped tusks, the use of which can only be conjectured. In the females these are but slightly developed.

It is quite clear, then, that we cannot place the Dinoceras in any order of living mammals. It is what palæontologists call a "generalised type;" that is to say, it presents certain characters seen in several groups of living quadrupeds, and not any of those

A LARGE EXTINCT MAMMAL, TINOCERAS INGENS.
From North America. Length about 12 feet (without the tail).

PLATE XIV.

elaborated or highly developed parts which we see in such animals to-day. Thus the proboscis of the elephant is a greatly elongated nose; in other words, the elephant is highly " specialised " in that direction, whereas our Dinoceras had no proboscis, or only a very slight one.

Again, the six remarkable bony protuberances of the skull served to some extent as horns, and probably were covered with thick bosses of skin, and did not support true horns

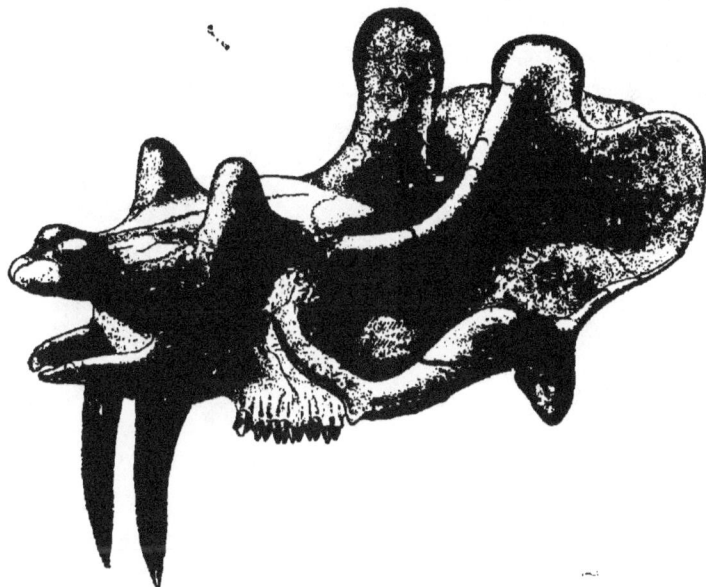

FIG. 43.—Skull of *Dinoceras mirabile.* (After Marsh.)

like those of our modern oxen and other ruminants. Speaking of these protuberances, Professor Marsh says, " None of the covering of these elevations, or horn-cores, has, of course, been preserved; yet a fortunate discovery may perhaps reveal their nature by the form of a natural cast, as the eyeball of the Oreodon is sometimes thus clearly indicated in the fine Miocene matrix which envelops these animals." It looks rather as if we have here an early stage in the evolution of horns,

and it may be that in the course of subsequent ages such prominences as those developed into true " horn cores," such as sheep
or goats have, while the thick bosses of skin that covered them
slowly developed into the true horns that are attached to these
cores.   If this is so, then we have here another instance of a
" generalised " structure.   Again, the limbs with their five toes
tell us at once that the creature's place in Nature is outside of
those two great groups of modern ungulates, or hoofed quadrupeds, the odd-toed and the even-toed, represented on the one
hand by the horse, rhinoceros, and tapir, on the other by the pig,
camel, deer, ox, and many other forms.   Probably the two groups

FIG. 44.—Cast of brain-cavity of *Dinoceras mirabile.*   (After Marsh.)

had not at this early period branched off from the primitive
ungulate stock with five toes in each foot, of which the elephant
is a living descendant, and from which also the Dinoceras must
have come.

The limbs were strong and massive, but the brain was
remarkably small, so that our Dinoceras cannot be credited
with any high degree of intelligence : and here again we see
an absence of "specialisation" compared with the sagacious
elephant.   Professor Marsh has taken casts of its brain-cavity
(see Fig. 44).   These casts show that the brain was smaller (in
proportion to the size of the animal) than in any other mammal,
whether living or extinct—and even less than in some reptiles !

In fact, it was a decidedly reptilian kind of brain. Perhaps it may seem hardly credible, but so small was the brain of Dinoceras mirabile, that it could have been pulled through the apertures (neural canals) of all the neck vertebræ! In certain marsupials of the present day we find an approach to this kind of brain. It seems to be an established fact, according to Professor Marsh, that all the Eocene or earlier Tertiary mammals had small brains. His researches among fossil mammals have led him to the important conclusion that, as time went on, the brains of mammals grew larger; and thus he has been able to establish his law of brain-growth during the Tertiary period, a law which appears to be plainly recorded in the fossil skulls of succeeding races of ancient mammals. The importance of a discovery such as this cannot fail to strike the imagination of even the most unlearned in geology as being singularly suggestive and instructive. It is not difficult to picture these dull, heavy, slow-moving creatures haunting the forests and palm jungles around the margin of the great Eocene lake, into the waters of which their carcases from time to time found their way—perhaps swept down by floods. No footprints have been discovered as yet.

The Dinocerata were very abundant for a long time during the middle of the Eocene period. The position of their remains suggests that they lived together in herds, as cattle do now, and they probably found an abundance of food in the shape of succulent vegetation round the great lake. Geological evidence points to their sudden extinction before the close of the Eocene period; but it is difficult to understand this. Professor Marsh thinks that from their sluggish nature they were incapable of adapting themselves with sufficient rapidity and readiness to new conditions, such as may have been brought about by geographical changes. It must be admitted, however, that the geological record in this region does not give evidence of any sudden change. Possibly they may only have migrated to some other region, where their remains have not yet been discovered, or where, for various reasons, their skeletons were not preserved.   In this Eocene lake, where sediments went on being quietly deposited for a long time, we have the most favourable conditions for the preservation of the different forms of life that flourished round its borders.

In the museum at Yale College are collected the spoils of numerous expeditions to the West, and the many tons of bones lying there are believed to represent the remains of no less than two hundred individuals of the Dinocerata.   So perfectly have these bones been preserved by Nature that, even if the creatures had been living now, the material for studying their skeletons could hardly be more complete.   Professor Marsh recognises three distinct types in this strange group of quadrupeds, on each of which a genus has been founded.   The first and oldest form is the Uintatherium, which takes its name from the Uinta Mountains. This, as might be expected, is the most primitive or least specialised form, and comes from lower strata.   The most highly developed or specialised form is the Tinoceras, and this is found at the highest geological level or "horizon."

Between these two extremes, and from an intermediate horizon, comes the Dinoceras,[1] so that in tracing these animals through the strata in which they occur the geologist finds that he is following for a while the course of their evolution.   Doubtless there were many slight differences presented by the members of this group, but at present it has not been found possible to determine the number of species, although about thirty forms more or less distinct have been recognised.   Professor Marsh says that the specimen of the skull of Dinoceras mirabile, on which the whole order Dinocerata was founded, is, fortunately, in a very perfect state of preservation, and that it belonged to a fully adult animal.   Moreover, it was embedded in so soft a

[1] The Dinoceras of Marsh is the same form as Eobasileus of Cope.   Uintatherium was discovered by Leidy.

matrix that the brain-cavity and the openings leading from it could be worked out without difficulty. In removing the skull from the rock, on the high and almost inaccessible cliff where it was found, two or three important fragments were lost; but Professor Marsh, after a laborious search, recovered them from the bottom of a deep ravine, where they had been washed down and covered up.

It is about twenty-two years since the wonderful forms of life sealed up within these Eocene lake-deposits first became known to science. Long before then, however, the wandering Indian had been accustomed to seeing strange-looking skulls and skeletons that peeped out upon him from the sides of cañons and hills, as the rocks that enclosed them crumbled away under the influence of atmospheric agents of change—the ceaseless working of wind, rain, heat, and cold. To his untrained mind no other explanation suggested itself than the idea that these were the bones of his ancestors, which it would be highly impious to disturb. *Requiescant in pace !* So he left them in peace. Perhaps he believed in a former race of human giants; if so, these would be their bones. Long before Professor Marsh's expeditions, the earliest squatters, trappers, and others used to bring back news of marvellous monsters grinning from the ledges of rock beneath which they camped. At last these tales attracted the notice of some enthusiastic naturalists in the eastern States. Professor Leidy obtained a number of bones, from which he was able to bring to light an extinct creature at that time unknown to science, namely, the Uintatherium. Professor Cope also described some extinct animals disinterred by himself from the same region.

But our knowledge of the Dinocerata is chiefly due to Professor Marsh, who has despatched one expedition after another, and who, after many years of laborious research both in the western deserts and in his wonderful collection at Yale College, has published a splendid monograph on the subject.

No trouble and no expense have been spared in order to obtain material for this great work, and all geologists must feel grateful to Professor Marsh for so liberally devoting his time and his private resources in order to advance the science of Palæontology.

The region in which the remains occur of the remarkable group of extinct animals now under consideration, has a peculiar scenery of its own, unlike anything in Europe. The following graphic description of its features is from the pen of Sir Archibald Geikie :—[1]

"On the high plateau that lies to the west of the Rocky Mountains, along the southern borders of the Wyoming territory, the traveller moving westwards begins to enter on peculiar scenery. Bare, treeless wastes of naked stone, rising here and there into terraced ledges and strange tower-like prominences, or sinking into hollows where the water gathers in salt or bitter pools. Under the cloudless sky, and in the clear dry atmosphere, the extraordinary colouring of these landscapes forms, perhaps, their weirdest feature. Bars of deep red alternate with strips of orange, now deepening into sombre browns, now blazing out again into vermilion, with belts of lilac, buff, pale green, and white. And everywhere the colours run in almost horizontal bands, running across hollows and river-gorges for mile after mile through this rocky desert. The parallel strips of colour mark the strata that cover all this wide plateau country. They are the tints characteristic of an enormous accumulation of sedimentary rocks, that mark the site of a vast Eocene lake, or succession of lakes, on what is now nearly the crest of the continent."

In this strange region the flat-topped hills, table-lands, or terraces, as they are variously named, seen from lower levels, are usually called "buttes," especially when they are of limited extent. This name is of French origin, and signifies a bank of

[1] *Nature,* vol. xxxii. p. 97.

earth or rising ground. It is also applied in a limited sense to
the more prominent irregularities of the deeply sculptured slopes
of the larger terraces. These buttes, therefore, vary in extent, from
a mere mound rising slightly above the level of the plains to hills
of varied configuration reaching to the level of the broader buttes
or terraces.

The *débris* resulting from the continual wearing away, or
demolition of these buttes and terraces, now lies spread out on
the plains below. From the lower plains the smaller terraces ·
appear like vast earth-work fortifications, and when not too much
cut up by erosion, remind one of long railway embankments.
But in many cases the terraces are so much cut up by narrow
ravines that they appear as great groups of naked buttes rising
from the midst of the plain. Nothing can be more desolate in
appearance than some of these vast assemblages of crumbling
buttes, destitute of vegetation, and traversed by ravines, in
which the watercourses in midsummer are almost all dried up.
To these assemblages of naked buttes, often worn into castellated
and fantastic forms, and extending through miles and miles of
territory, the early Cañadian *voyageurs* gave the name *Mauvais* ʃ
Terres. They occur in many localities of the Tertiary formations
west of the Mississippi River. Professor Leidy, who with two
friends made an expedition in search of fossils to Dry Creek Cañon
in this region of the "Bad Lands," about forty miles to the south-
east of Fort Bridger (Wyoming), thus describes his impressions :—

"On descending the butte to the east of our camp, I found
before me another valley, a treeless barren plain, probably ten
miles in width. From the far side of this valley butte after butte
arose and grouped themselves along the horizon, and looked
together in the distance like the huge fortified city of a giant
race, the utter desolation of the scene, the dried-up watercourses,
the absence of any moving object, the profound silence which
prevailed, produced a feeling that was positively oppressive.
When I thought of the buttes beneath our feet, with their

entombed remains of multitudes of animals for ever extinct, and reflected upon the time when the country teemed with life, I truly felt that I was standing on the wreck of a former world."

These old lake-basins, in which so many forms of life have been sealed up, all lie between the Rocky Mountains on the east, and the Wasatch Range on the west, or along the high central plateau of the continent. As the mountains were slowly elevated, part of the old sea of the Cretaceous period (that sea in which the "sea-serpents" played so important a part) was enclosed and cut off from the ocean. Rivers began to pour their waters into it, so that the waters became less and less salt, until at last a fresh-water lake, or series of lakes, was formed. As the upward movement of this region continued these lakes were all the while receiving sedimentary materials, such as sand and mud, from the rivers, until finally they were filled up, but not until the sediments had formed a mass of strata over a mile in thickness. Thus we see how favourable were the conditions for a faithful record of Eocene life-history.

But another process was going on which helped to bring them to an end ; for they were being slowly drained by the rivers that flowed out of them, and these rivers kept on continually deepening their channels, so that we have dry land where the lakes once were. *Now* the region is over 6000 feet above the sea, and probably more than one-half of these fresh-water deposits have been washed away, mainly through the Colorado River. What is left of the Eocene strata forms the " Bad Lands." The same geological action that has cut up and carved out this region into buttes, cañons, cliffs, peaks, and columns of fantastic shapes, has also brought to light the extinct animals preserved in the rocks, much in the same way as an old burial-ground, if cut up by intersecting trenches, might be made to yield up the bones of those who for generations had been buried therein.

Professor Marsh first discovered remains of Dinocerata in 1870, while investigating this Eocene lake-basin, which had never before been explored. It was here, also, that he found the wonderful series of fossil horses by means of which he has been able to prove that our modern horse is descended from a small quadruped with five toes, and to show the different stages in its evolution. Here, also, were found old-fashioned types of carnivorous quadrupeds, of rodents, and of insectivorous creatures. But reptiles as well as quadrupeds flourished on the borders · of the old lake, for the remains were found of crocodiles, tortoises, lizards, and serpents; its waters, too, were well stocked with fish.

Everything here testifies to a long continuance of those conditions under which plant and animal life can flourish, namely, a warm climate, plenty of food, and freedom from those physical changes which, by altering the geographical features of a country, bring so many important consequences in their train. The geological record tells us that this happy state of things lasted all through the Eocene period, and until the fresh-water lakes had at last been drained away by their outflowing rivers.

In October, 1870, a later Eocene lake-basin was discovered by the same exploring party, and this Professor Marsh calls the Uinta basin, because it was situated south of the Uinta Mountains. "In the attempt to explore it," he says, "our party endured much hardship, and also were exposed to serious danger, since we had only a small escort of United States soldiers, and the region visited was one of the favourite resorts of the Uinta-Utes. These Indians were then, many of them, insolent and aggressive, and since have been openly hostile, at one time massacring a large body of Government troops sent against them. Two subsequent attempts . . . to explore this region met with little success."

This lower lake was of later (or upper) Eocene age, and the extinct animals preserved in its ancient bed appear to resemble more nearly those of the famous Paris basin, referred to in the beginning of this chapter, than any yet discovered in America. But the basin north of the Uinta Mountains, where alone the

Dinocerata had been found, offered so inviting a field that, in the spring of 1871, Professor Marsh began to explore it systematically. He organised an expedition, with an escort of U.S. soldiers, and the work continued during the whole season. In this way a large collection was secured. Explorations were continued in the spring of the following year, which resulted in the discovery of the type specimen of the Dinoceras mirabile. Another expedition was organised in 1873, also with an escort of soldiers, and a great many specimens were collected. These researches were continued during 1874, and again in 1875, with good results. Since then various small parties have been equipped and sent out by Professor Marsh to collect in the same region of the " Bad Lands ; " and, finally, during the entire season of 1882, the work was vigorously prosecuted under his direction, and afterwards under the auspices of the United States Geological Survey. This brief account of the difficulties and hardships encountered by Professor Marsh and his companions, for which we are indebted to his exhaustive monograph, will serve to give some idea of the nature of those labours, undertaken in the cause of Science, which he has brought to so successful an issue.

In the country east of the Rocky Mountains, including the states of Dakota, Nebraska, Wyoming, and part of Colorado, Professor Marsh has discovered the remains of yet another strange group of large quadrupeds. The best known of these is Brontops, of which the skeleton is seen in Fig. 45. These animals lived after the Dinocerata, namely, in the Miocene period, and were the largest American mammals of that period. They constitute a distinct family more nearly allied to the rhinoceros than to any other living form. The skeleton on which Fig. 45 is founded was the most complete of any yet discovered by Professor Marsh. Portions of it were exhumed at different times, but it was first found in 1874. Our artist has made the restoration seen in Plate XV. from this skeleton, as figured by Professor Marsh.

A HUGE EXTINCT MAMMAL FROM NORTH AMERICA. BRONTOPS ROBUSTUS.

Height 8 feet.

PLATE XV.

This strange group of creatures flourished in great numbers on the borders of an old lake of Miocene age.   The Brontops was a heavy massive animal, larger than any of the Dinocerata, with a length of twelve feet, not including the tail, and a height of eight feet.   The limbs are shorter than those of the elephant, which it nearly equalled in size.   As in the tapir, there were four toes to the front limbs, and three to the hind limbs.   Its skull was of a peculiar shape, shallow, and very large.   That of Brontops ingens is thirty-

FIG. 45.—Skeleton of Brontops robustus.   (After Marsh.)

six inches long, and twenty inches between the tips of the two horns, or protuberances.   The creature was probably provided with an elongated, flexible nose, like that of the tapir, but not longer, because the length of the neck shows that it could reach the ground without the aid of a trunk such as the elephant's.   It is doubtful if the two prominences on the front of the skull were provided with horns, for, if directed forwards, they would interfere with the animal when grazing.

M

CHAPTER XI.

" What a glorious privilege it would be, could we live back—were it but for an instant—into those ancient times when these extinct animals peopled the earth ! to see them all congregated together in one grand natural menagerie—these mastodons and elephants, so numerous in species, toiling their ponderous forms and trumpeting their march in countless herds through the swamps and reedy forests ! "—HUGH FALCONER.

IT is a far cry back, against the sun's path, from Wyoming and the flanks of the Rocky Mountains to the sacred Himalayas—the " abode of snow "—of Northern India. But if the reader will follow us to that country, we will endeavour to describe two or three out of many strange and now lost forms of life brought to light from the famous Sivalik Hills, on the southern border of the Himalayas for the knowledge of which Science is greatly indebted to a very distinguished palæontologist, the late Mr. Hugh Falconer. Together with his friend Captain Cautley (afterwards Sir Proby Cautley), he explored this region, and their joint arduous labours show that it was at one time inhabited by a very large and varied group of quadrupeds, together with many birds, reptiles, fishes, mollusca, and crustaceans.

In this region there lived, throughout a considerable part of the Tertiary period, elephants, of various species, whose skulls and bones were found in great numbers; mastodons (a closely allied form) ; and several species of hippopotamus, rhinoceros, and horse: among ruminants, species of the camel, the ox, the stag, and the antelope, together with a colossal creature unknown before, the Sivatherium, which has never been found elsewhere; a huge tortoise, and various species of carnivora, rodents, and apes.

With regard to the geography of the region, it appears that the continent of India, at an early period of the Tertiary era, was a large island, situated in a bight, or bay, formed by the Himalayas and the Hindo Koosh range. The valleys of the Ganges and Indus formed a long estuary, into which the drainage of the Himalayas poured its silt and alluvium. Later on, an upheaval took place, converting these straits into the plains of India, connecting them with the ancient island, and forming the existing continent. The large and varied forms whose remains now lie "sealed within the iron hills" then spread over the continent, from the Irrawaddi to the mouths of the Indus, two thousand miles; and north-west to the Jhelum, fifteen hundred miles. After a long interval of repose, another great upheaval took place, which threw up a strip of the plains of India, crumpled and ridged it up to form the Sivalik Hills, and at the same time increased the elevation of the Himalayas by many thousands of feet.

It would be easy to show that such events as these must have been followed by changes in climate, for the climate of a region depends largely on its physical features—the proportion of land and water, the presence of hills and mountain ranges, and their height; and it is considered probable that the physical changes above mentioned helped to bring about the extinction of this most interesting and ancient fauna. Throughout the latter part of the Tertiary era it is well known to geologists that the climate of Europe was becoming gradually colder, until at last a glacial period, or "Ice Age," was experienced, during which Northern Europe was subjected to an arctic climate, and the great ice-sheet seems to have been slowly retiring and melting away in the early part of the Stone Age. But in India there has been no such decrease in temperature, and it enjoyed in Tertiary times as warm a climate as it now has, so that both animal and vegetable life continued to flourish vigorously.

By the Sivalik (or Sewalik) Hills is meant that range of lower elevations which stretches along the south-west foot of the Himalayas, for the greater portion of their extent from the Indus to the Brahmapootra, where those rivers respectively debouche from the hills into the plains of India. It extends for nearly a thousand miles, and it appears to have been entirely built up of alluvial *débris*, washed down from the Himalayas into that sea which we have already referred to as having once separated the plains of India from the great range now forming its northern boundary. The strata thus formed were subsequently upheaved to form the Sivalik Hills. Thus we see that one mountain range may help to form another one running parallel to itself. The name is derived from Siva, or Mahadeo, the Hindoo god; these hills, as well as the Himalayas, being connected in Hindoo mythology in various ways with the history of Siva.

Dr. Falconer and Captain Cautley soon found that they had "struck oil" in the Sivalik Hills, or, in other words, had come upon one of Nature's great graveyards, full of material most valuable to the palæontologist—one which, extending for hundreds of miles, might perhaps prove to be as rich in relics of the world's "lost creations" as the lake-basin in Wyoming, where Professor Marsh discovered his Dinocerata and other extinct types.

Let us give Dr. Falconer and Captain Cautley their due. They found themselves suddenly confronted with a perfect mine of wealth, in a far country, where the ordinary means resorted to by men of science for determining extinct types and species, by comparison with living forms, were not to be obtained, for there were no libraries and no museums of comparative anatomy in that remote quarter of India. But Dr. Falconer was not the man to be baffled by such drawbacks, which would have deterred and discouraged some men. He appealed to the living forms that abounded in the surrounding forests, rivers, and swamps, and

took toll of them to supply the want.   Nature herself became his
library and his museum.   Skeletons of all kinds were prepared ;
the extinct forms he collected were compared with their nearest
living allies, and a valuable series of "Memoirs" by himself and
Captain Cautley was the result.[1]

The Sivalik explorations soon attracted attention in Europe,
and in 1837 the Wollaston Medal, in duplicate, was awarded for
their discoveries to Dr. Falconer and Captain Cautley by the
Geological Society, the fountain of geological honours in England ;
while the value of the distinction was enhanced by the terms
in which the President, Sir Charles Lyell, was pleased to an-
nounce the award.   This is what he said : " When Captain Cautley
and Dr. Falconer first discovered these remarkable remains, their
curiosity was awakened, and they felt convinced of their great
scientific value ; but they were not versed in fossil osteology [the
study of bones], and, being stationed on the remote confines of
our Indian possessions, they were far distant from any living
authorities or books on comparative anatomy to which they could
refer.   The manner in which they overcame these disadvantages,
and the enthusiasm with which they continued for years to prosecute
their researches, when thus isolated from the scientific world, are
truly admirable.   Dr. Royle has permitted me to read a part
of their correspondence with him, when they were exploring the
Sivalik Mountains, and I can bear witness to their extraordinary
energy and perseverance.   From time to time they earnestly
requested that Cuvier's works might be sent out to them, and
expressed their disappointment when, from various accidents,
these volumes failed to arrive.   The delay, perhaps, was fortunate ; for, being thrown entirely upon their own resources, they
soon found a museum of comparative anatomy in the surrounding
plains, hills, and jungles, where they slew the wild tigers, buffaloes,
antelopes, and other Indian quadrupeds, of which they preserved

[1] These appeared in the *Asiatic Researches*, the *Journal of the Asiatic Society
of Bengal*, and in the *Geological Transactions* of the London Geological Society.

the skeletons, besides obtaining specimens of all the reptiles which inhabited that region.   They were compelled to see and think for themselves, while comparing and discriminating the different recent and fossil bones, and reasoning on the laws of comparative osteology, till at length they were fully prepared to appreciate the lessons which they were taught by the works of Cuvier."

In 1840 Captain Cautley presented his vast collection, the result of ten years' unremitting labour and great personal outlay, to the British Museum, the Geological Society having declined to accept it, as it was beyond their means of accommodation.   Its extent and value may be estimated from the fact that it filled 214 large chests, the average weight of each of which amounted to 4 cwt., and that the charges on its transmission to England alone, which were defrayed by the Government of India, amounted to £602.   Dr. Falconer's selected collection was divided between the India House and the British Museum ; the greater part was presented to the former, but a large number of unique or choice specimens, required to fill up blanks, were presented to the latter. The greater part of the specimens in the British Museum were still unarranged and embedded in their matrix.   In 1844 a memorial was presented to the Court of Directors of the Honourable East India Company, pointing out the desirability of having the specimens in the national collection prepared, arranged, and displayed, and also of publishing an illustrated work, which would convey to men of science in both hemispheres a knowledge of the contents of the Sivalik Hills, and suggesting Dr. Falconer as the person most fitted to superintend the work.   The Government of the time, under Sir Robert Peel, made a grant of £1000 to enable the collection to be exhibited in the British Museum, and Dr. Falconer was entrusted with the work.   Besides this, a large illustrated work, entitled *Fauna Antiqua Sivalensis*, was begun, but owing to the demands upon Dr. Falconer's time, and his subsequent death, this work was not completed, although nine out of the twelve parts originally contemplated were finished.

The great Indian collection of fossils, mainly the gift of Sir Proby Cautley (the specimens of which, stupendous in their size, and in fine preservation, were prepared, identified, and arranged by Dr. Falconer), has long constituted one of the chief ornaments of the collection at the British Museum—now removed to the Natural History Museum, Cromwell Road, South Kensington.

Other collections of fossils from the Sivalik Hills have been presented to the Museum of Edinburgh University by Colonel Colvin, and to the Oxford University by Mr. Walter Ewer. When it is remembered that these collections have since been increased tenfold, and that the remains were either excavated or found in the *débris* of cliffs, and that the explored surface bears a very small proportion to that which has not yet been investigated, one may conceive how prodigious must have been the number of animals that lived together in the former plains of India, even when every allowance is made for the bones having accumulated during many successive generations in the Sivalik strata.

From this large and important collection we select two of special interest for brief notice here, namely, the Sivatherium,[1] and an immense tortoise known as the Colossochelys.

The first of these monsters was a remarkable form of animal, unlike anything living. In size it surpassed the largest rhinoceros, and was bigger than any living ruminant. Altogether, it was one of the most remarkable forms of life yet detected in the more recent strata. It had two pairs of horns on its head—two short and quite simple ones in front, and two larger ones, more or less expanded, behind them. From the character of these long horn-cores, which are prolongations of the skull, it may be concluded that the Sivatherium was a gigantic ruminant with four horns. A cast of the original skull, with the horn-cores restored from actual parts, in the collection and elsewhere, has been placed on a stand in the centre of the long gallery of fossil vertebrates at

[1] From *Siva*, the Hindoo god ; and Greek, *therion*, a beast.

South Kensington (Stand I) near to the case containing the skull and other portions of the skeleton (see Fig. 46). There is also hanging on the wall near, a clever painting by Berjeau, representing the creature as it may have appeared when alive. The entire skeleton, partly restored, is shown in Fig. 47, with a conjectural outline of the body. A hornless skull of a nearly allied animal from the same strata and locality is placed with that of the Sivatherium, and was considered by Dr. Falconer and

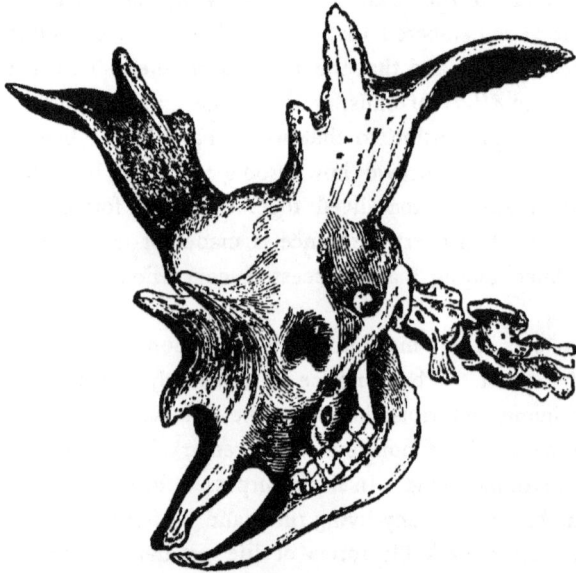

FIG. 46.—Skull of *Sivatherium giganteum*, from the Sivalik Hills, Northern India.

others to be the skull of the hornless female (also represented as such in the above picture referred to); but is now, by more recent writers, regarded as a separate genus, viz. the Helladotherium, so named because the remains were first discovered at Pikermi, near Athens, Greece (ancient Hellas). (See Plate XVI.)

In the Sivatherium we have a new type which seems to connect together two families at the present time well marked off

A GIGANTIC HOOFED MAMMAL, SIVATHERIUM GIGANTEUM.

From the Sivalik Hills, Northern India. An allied form, *Helladotherium*, is seen on the left.

PLATE XVI.

from each other, namely, the giraffe and the antelope. Its teeth resemble those of the former animal, while in its four horns it resembles a certain antelope (Antilope quadricornis). The head in certain respects shows resemblances to that of the ox, but the upper lip must have been prolonged into a short proboscis, or trunk, like that of the tapir. The form and proportions of the jaw agree closely with the corresponding parts of a buffalo. But no known ruminant, fossil or existing, has a jaw of such large

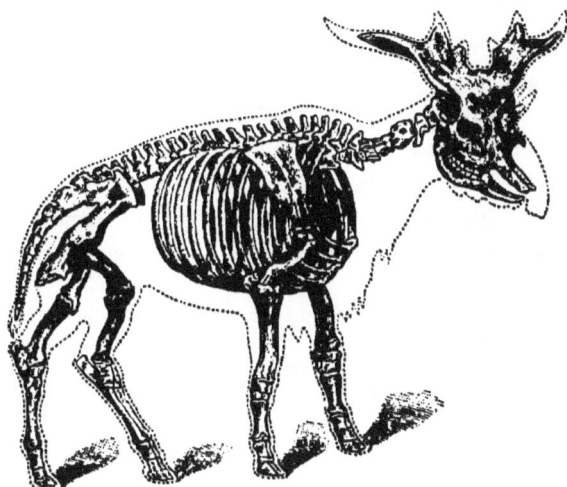

FIG. 47.—Skeleton of *Sivatherium giganteum.*

size, the average dimensions being more than double those of a buffalo. The skull is the best known part of the animal, but Captain Cautley came across some of the bones of the limbs.

The Colossochelys atlas,[1] or gigantic fossil tortoise of India, supplies a fit representative of the tortoise which sustained the elephant and the infant world in the fables of the Pythagorean and Hindoo cosmogonies. It is highly interesting to trace back to its probable source a matter of belief like this, so widely con-

[1] Greek, *Colossos,* Colossus, and *chelus,* tortoise. Atlas was supposed to sustain the world on his shoulders.

connected with the speculations of an early period of the human race.

The carapace, or buckler, of the shell of this crawling monster is similar in general form to the large land-tortoises of the present day.[1] The shell is estimated to have been at least six feet long. The limbs were probably similar to those of a modern land-tortoise, and the limb-bones are of huge size—a single humerus, or arm-bone, measuring 28 inches. Probably the foot was as large as that of a rhinoceros. A restored cast of a young individual stands at the West end of the fossil reptile gallery, South Kensington (Stand Z on plan). Length of the shield, 10 feet[2] (see Fig. 48).

The first fossil remains of this colossal tortoise were discovered by Dr. Falconer and Captain Cautley in 1835, in the Tertiary strata of the Sivalik Hills. At the period when it was living—probably the Pliocene—there was great abundance and variety of life on the scene, for its remains were found to be associated with

[1] Giant tortoises of the present day live on islands—where they have escaped competition with large carnivora and other foes—such as the Aldabra group, N.W. of Madagascar, in the Mascarenes, which comprise Mauritius and Rodriguez ; and the Galapagos, or "Tortoise Islands," off the coast of South America. When Mr. Darwin visited the latter islands he saw the relics, as it were, of a family of huge tortoises, which lived there in abundance a few years before, and was able to verify many interesting facts which had been recorded by Porter in 1813, who stated that some of those captured by him weighed from 300 to 400 lbs., and that on one island they were 5½ feet long. Those of one island differed from those of another. Some had long necks. After Mr. Darwin's visit the process of extermination went on. At the present time it is most probable that the gigantic tortoises are very rare where formerly they were so abundant. One of these great tortoises is that of Abingdon Island, in the Galapagos Archipelago, of which there is a fine stuffed specimen in the Natural History Museum (Reptile Gallery). It has a very long neck, and a small flat-topped head with a short snout. It weighed originally 201 lbs. The Indian tortoises of the present day are not of large size. See the fine specimens in the Natural History Museum—Reptile Gallery (left wing of the building).

[2] Dr. Falconer's estimate was much too great, so that this model is too large. Mr. Lydekker prefers to drop the generic term Colossochelys, and call it Testudo Atlas. In length it was only one-third greater than Testudo elephantina of the Galapagos Islands.

those of many great quadrupeds, such as the elephant, mastodon, rhinoceros, horse, camel, giraffe, sivatherium, and many other mammals. The Sivalik fauna also included a great number of reptiles, such as crocodiles, lizards, and snakes.

The greater part of the remains of the Colossochelys atlas were

FIG. 48.—Restored figure of gigantic tortoise, *Colossochelys atlas*, from the Sivalik Hills, Northern India.

collected during a period of eight or nine years, along a range of about a hundred miles of hilly country. Consequently, they belong to a large number of individuals, varying in size and age. They were met with in crushed fragments, contained in upheaved strata, which have undergone considerable disturbance, so that it is improbable that an entire uncrushed specimen will ever be found. When the first fragments, in huge shapeless masses, were found by the discoverers, they were utterly at a loss what to make of them, and for many months could do nothing more than look upon them in bewildered and nearly hopeless admiration. But no sooner was the clue found to a single specimen than every fragment moved into its place so as to form a consistent whole.

It is not possible at present to say, with any degree of certainty, whether this colossal tortoise survived into the human period ; but at least there is no evidence against the idea, and Dr. Falconer shows it is quite possible that the frequent allusions to a gigantic tortoise in Hindoo and other mythologies are to be explained on the supposition that the creature was seen by the men of a prehistoric age. Other species of tortoises and turtles that were coeval with the Colossochelys have lived on to the present day. So have other reptiles, for some of the crocodiles now living in India appear to be identical with the forms dug out of the Sivalik Hills. In the absence of direct geological evidence, we must fall back on traditions.

Now, there are traditions connected with the speculations of nearly all Eastern nations with regard to the world (cosmogonies) that refer to a tortoise of such gigantic size as to be associated with the elephant, in their fables. The question therefore arises —Was this tortoise a creature of the imagination, or was the idea of it drawn from a living reality? Besides a tradition current among the Iroquois Indians of North America, referring to the important share which the tortoise had in the formation of the earth, there are several cases in ancient history bearing on the same point. Thus, we find in the Pythagorean doctrine the infant world represented as having been placed on the back of an elephant, which was sustained on a huge tortoise. Greek and Hindoo mythologies were undoubtedly related to each other, and accordingly we find, in the Hindoo accounts of the second

Avatar of Vishnoo, that the ocean is said to have been churned by means of the mountain placed on the back of the king of the tortoises, and the serpent Asokee used as the churning-rope. Again, Vishnoo was said to have assumed the form of the tortoise, and to have sustained the created world on his back to make it stable. This fable has taken such a firm hold of the Hindoos, that to this day they believe the world rests on the back of a tortoise (see Fig. 49). In the narratives of the feasts of the bird-demigod, Garūda, the tortoise again figures largely,

FIG. 49.—The elephant victorious over the tortoise, supporting the world, and unfolding the mysteries of the *Fauna Sivalensis.* From a sketch in pencil in one of Dr. Falconer's note-books, by the late Professor Edward Forbes.

and Gurūda is said on one occasion to have appeased his hunger at a certain lake where an elephant and a tortoise were fighting.

These three instances, in each of which there is a distinct reference to a gigantic form of tortoise, comparable in size with the elephant, suggest the question whether we are to regard the idea as a mere fiction of the imagination, like the Minotaur or the Chimæra, or as founded on a living tortoise. Dr. Falconer points out that it seems unlikely that such fables could have been suggested by any of the small species of tortoises now living in India, and consequently is inclined to think that the monster was seen by man many centuries ago, long before he began to write history. We have already alluded to the large number of mammalian forms of life that were contemporary with the Sivatherium and Colossochelys, but if we examine this old Sivalik fauna we find it presents several very interesting features. In the first place, it exhibits a wonderful richness and variety of forms, compared to the living fauna of India. Take the pachy-dermata, for instance—an old order established by Cuvier to include the rhinoceros, hippopotamus, elephant, etc.—and we find there were, in the period under consideration, about five times the number of species now known in India. Elephants and mastodons, too, of various species abounded. So it is with the ruminants ; besides a large number of species allied to those now living, such as the ox, buffalo, bison, deer, antelope, musk-deer, and others, there were giraffes and camels, as well as the strange Sivatherium. And so it is with the other orders, such as carnivora, rodents, insectivora, etc.

Secondly, this great and varied fauna of the past shows a striking resemblance to that of India at the present day. Darwin found the same resemblance in South America ; and now it is accepted as a general law, that the living fauna of a country resembles its extinct fauna, especially that of the latest geological period. Dr. Falconer found that India's living fauna is but, as it were, a remnant of that which it once possessed.

Thirdly, this extinct Sivalik fauna presents a singular mixture of old and new forms. And lastly, it points to a very different geographical distribution of animals. Thus the giraffe, the hippo-potamus, and the ostrich are *now* confined to Africa. Facts such as these serve to throw light on the geography of the past ; but we cannot stay to enlarge on that subject here.

Much might be said about the fossil elephants and mastodons from the Sivalik Hills, so fully described by Dr. Falconer, but since

chapters xiii. and xiv. deal with elephants, we must reserve
our remarks till then, only alluding here to one striking form
from the Sivalik Hills, namely, the Elephas ganesa, the tusks of
which were more than ten feet in length, and much less curved
than those of the mammoth. A very fine specimen of the head
and tusks may be seen in the gallery of fossil mammals in the
Natural History Museum (Gallery I, Stand D).

With the following eloquent passage from Dr. Falconer's
" Memoirs," we take leave of the remarkable Sivalik fauna, hoping
that future geologists will endeavour to follow his example and
bring to light yet other "lost creations" from that region, so
rich in fossils, yet comparatively unexplored. Would that the
English Government could see their way to follow the example
of the United States, and send out a scientific expedition to
explore this wonderful region! There can be no doubt that a
rich harvest lies waiting there to be reaped.

" What a glorious privilege it would be, could we live back—
were it but for an instant—into those ancient times when these
extinct animals peopled the earth! to see them all congregated
together in one grand natural menagerie—these mastodons and
elephants, so numerous in species, toiling their ponderous forms
and trumpeting their march in countless herds through the swamps
and reedy forests! to view the giant Sivatherium, armed in front
with four horns, spurning the timidity of his race, and, ruminant
though he be, proud in his strength, and bellowing his sturdy
career in defiance of all aggression! And then the graceful
giraffes, flitting their shadowy forms like spectres through the
trees, mixed with troops of large as well as pigmy horses, and
camels, antelopes, and deer! And then, last of all, by way of
contrast, to contemplate the colossus of the tortoise race, heaving
his unwieldy frame, and stamping his toilsome march along
plains which hardly look over strong to sustain him!

" Assuredly it would be a heart-stirring sight to behold! But
although we may not actually enjoy the effect of the living

pageant, a still higher order of privilege is vouchsafed to us. We have only to light the torch of philosophy, to seize the clue of induction, and, like the Prophet Ezekiel in the vision, to proceed into the valley of death, when the graves open before us and render forth their contents ; the dry and fragmented bones run together, each bone to his bone ; the sinews are laid over, the flesh is brought on, the skin covers all, and the past existence —*to the mind's eye*—starts again into being, decked out in all the lineaments of life. 'He who calls that which hath vanished back again into being, enjoys a bliss like that of creating.' Such were the words of the philosophical Niebuhr, when attempting to fill up the blanks in the fragmentary records of the ancient Romans, whose period in relation to past time dates but as of yesterday. How much more highly privileged, then, are we, who can recall, as it were, the beings of countless remote ages, when man was not yet dreamed of ! not only this, but if we use discreetly the lights which have been given to us, we may invoke the spirit of the winds, and learn how *they* were tempered to suit the natures of these extinct beings."

CHAPTER XII.

" Injecta monstris terra dolet suis."

HORACE, *Odes,* book iii.

IT would have been strange, considering how much we owe to North America, had the great South American continent not enriched our knowledge of past forms of life on the globe. But such is not the case. The honours are, as it were, divided, although it must be admitted that the North American extinct forms at present known are far more numerous. There are, however, two or three " Extinct Monsters " of very great interest which once had a home in South America—in that strange region of the Pampas, where the naturalist of the present day finds so much to excite his interest. Of these the present chapter treats.

The Megatherium [1] (Cuvier) was a gigantic mammal allied to sloths and ant-eaters, and perhaps to the armadillos. In its skull and teeth this colossus of the past resembled the sloths, in its limbs and backbone it resembled the ant-eaters, while in size it surpassed the largest rhinoceros (Plate XVII.). The famous, but imperfect, specimen at Madrid was for a long time the principal if not the only source of information with regard to this extinct genus, and for nearly a century it remained unique.

Later on, however, the zeal and energy of Sir Woodbine Parish, his late Majesty's *chargé-d'affaires* at Buenos Ayres, greatly helped to augment the materials for arriving at a just conclusion

[1] Greek—*megas,* great ; *therion,* beast.

N

with regard to its proper place in the animal kingdom. According to one writer, Spain formerly possessed considerable parts of three different skeletons. The first and most complete is that which is preserved in the royal cabinet at Madrid. This was sent over in 1789, by the Marquis of Loreto, Viceroy of Buenos Ayres, with a notice stating that it was found on the banks of the river Luxan. In 1795 a second specimen arrived from Lima, and other portions, probably not very considerable, were in the possession of Father Fernando Scio, to whom they had been presented by a lady from Paraguay. But two German doctors, Messrs. Pander and D'Alton, who published in 1821 a beautiful monograph on the subject, state that they were unable in 1818 to find any traces of either the Lima specimen or that which had belonged to Fernando Scio.

The remains collected by Sir Woodbine Parish were discovered in the river Salado, which runs through the flat alluvial plains (Pampas) to the south of the city of Buenos Ayres, after a succession of three unusually dry seasons, "which lowered the waters in an extraordinary degree, and exposed parts of the pelvis to view as it stood upright in the bottom of the river."[1]

This and other parts having been carried to Buenos Ayres by the country people, were placed at the disposal of Sir Woodbine Parish by Don Hilario Sosa, the owner of the property on which the bones were found. A further inquiry was instituted by Sir Woodbine; and on his application, the governor granted assistance, the result of which was the discovery of the remains of two other skeletons on his Excellency's properties, at no great distance from the place where the first had been found. It was in the year 1832 that Sir Woodbine Parish sent his valuable collection of bones from Buenos Ayres, and presented them to the Royal College of Surgeons. These specimens formed the subject of

---

[1] " Some Account of the Remains of the *Megatherium* sent to England from Buenos Ayres, by Woodbine Parish, Jun., Esq., F.R.S.," by Wm. Clift, Esq., F.R.S., *Geological Transactions*, second series, vol. iii. p. 437.

CAST OF A SKELETON OF MEGATHERIUM AMERICANUM.

PLATE XVII.   Set up in the Natural History Museum.

Mr. Clift's memoir above quoted. But even then the materials were not complete for a thorough knowledge of the bony framework of the Megatherium, and it was not till 1845, when more remains (discovered near Luxan, 1837) reached this country, that Professor Owen was able to clear up one or two doubtful details. These were purchased by the trustees of the British Museum, and casts of the bones were taken. Among the various writings by learned men on the subject, Professor Owen's masterly description stands pre-eminent; indeed, he was the only one to solve the riddle, to thoroughly explain the structure of this giant sloth, and to show how its food was obtained.[1] Neither Cuvier, nor the German doctors, nor Mr. Clift had succeeded in so doing.

In the Natural History Museum (Stand O, Gallery No. 2 on plan) is a cast representing the animal nearly erect, and grasping a tree. This magnificent cast (see Plate XVII.) represents an animal eighteen feet in length, and its bones are more massive than those of the elephant. For instance, the thigh-bone is nearly thrice the thickness of the same bone in the largest of existing elephants, the circumference being equal to the entire length. To a comparative anatomist several striking indications of great strength present themselves; thus, not only the very forms of the bones themselves mean strength, but their surfaces, ridges, and crests are everywhere made rough for the firm attachment of powerful muscles and tendons. In the fore part of the body the skeleton is *comparatively* slender, but the hind quarters show enormous strength and weight combined. The tail, also, is very powerful and massive. The fore limbs are long, and evidently constructed for the exertion of great force. How this force was applied we shall see presently. In both sets of limbs we notice

---

[1] His views are expounded in his *Memoir on the Megatherium, or Giant Ground Sloth of America*, 1861, which is beautifully illustrated. The Royal Society gave £100 (part of a Government grant of £1000) to enable Professor Owen to carry out this important work.

powerful claws, such as might be used for scratching up the
ground near the roots of a tree, and it was at one time thought
that this was the way in which the creature obtained its leafy food,
namely, by digging up trees by the roots and then devouring the
leaves. But Professor Owen had another explanation.

As in the living sloths and armadillos (edentata [1]), there are no
teeth in the fore part of the jaw. The molar teeth, of which there
are five on each side of the upper jaw, and four in the lower, are
hollow prismatic cylinders, straight, seven to nine inches long,
and implanted in deep sockets. There are no other teeth, but
these are composed of different substances, and so arranged that,
as the tooth wears, the surface always presents a pair of transverse ridges, thus producing a dental apparatus well suited for
grinding up vegetable food. In the elephants, which live on
similar food, the grinding is effected by great molar teeth, which
are replaced by new ones as the old ones are worn away. In the
Megatherium, however, only *one* set of teeth was provided; but
these, by constant upward growth, and continual addition of new
matter beneath, lasted as long as the animal lived, and never
needed to be renewed.

On looking at the model so skilfully set up at South Kensington,
and especially at the front part of the skull, it will be seen that
the snout and lips must have been somewhat elongated, possibly
into a slight proboscis like that of the tapir. The specimens of
the lower jaw in the wall-case close by show that it was much
prolonged and grooved. This fact must be interpreted to mean
that the creature possessed a long and powerful tongue, aided by
which it could, like the giraffe, strip off the small branches of the
trees which it had broken or bent down within its reach.

A bony shield (or carapace) of a great armadillo was found
with one of the specimens described by Mr. Clift, and Buckland
and others thought it belonged to the Megatherium; but Owen

---

[1] This word, which means *toothless*, is misleading. All the edentata, however, agree in having no front, or incisor, teeth.

afterwards showed, by most clear and convincing reasoning from the skeleton, that the Megatherium could not have been protected as armadillos are, by such a shield (see p. 190).

And now we come to the question how it obtained its food. The idea of digging round trees with its claws in order to uproot them, must be partly, if not entirely, given up; for Professor Owen has proved, by a masterly piece of reasoning, that this cumbrous creature, instead of climbing up trees as modern sloths do, actually pulled down the tree bodily, or broke it short off above the ground by a *tour de force*, and, in order to do so, sat up on its huge haunches and tail as on a tripod, while it grasped the trunk in its long powerful arms! Marvellous as this may seem, it can be shown that every detail in its skeleton agrees with the idea. Of course there would be limits to possibilities in this direction, and the larger trees of the period must have been proof against any such Samson-like attempts on the part of the Megatherium; but when the trunk was too big, doubtless it pulled down some of the lower branches. Plate XVIII. is a restoration, by our artist, of the South Kensington skeleton.

Speaking of the extinct sloths of South America, Mr. Darwin thus describes Professor Owen's remarkable discovery: "The habits of these Megatheroid animals were a complete puzzle to naturalists until Professor Owen solved the problem with remarkable ingenuity. Their teeth indicate by their simple structure that these animals . . . lived on vegetable food, and probably on the leaves and small twigs of trees; their ponderous forms and great strong curved claws seem so little adapted for locomotion, that some eminent naturalists believed that, like sloths, to which they are intimately related, they subsisted by climbing, back downwards, on trees, and feeding on the leaves. It was a bold, not to say preposterous, idea to conceive even antediluvian trees with branches strong enough to bear animals as large as elephants. Professor Owen, with far more probability, believes that, instead of climbing on trees, they pulled the branches down to them, and tore up the smaller ones by the roots, and so fed on the leaves. The colossal breadth and weight of their hinder quarters, which can hardly be imagined without having been seen, become, on this view, of obvious service instead of being an encumbrance; their apparent clumsiness disappears. With their great tails and huge heels firmly fixed like a tripod in the ground, they could freely exert the full force of their most powerful arms and great claws."[1]

To this we may add Dean Buckland's description,[2] "His entire frame was an apparatus of colossal mechanism, adapted exactly to the work it had to do; strong and ponderous in proportion as this work was heavy, and calculated to be the vehicle of life and enjoyment to a gigantic race of quadrupeds, which, though they have ceased to be counted among the living inhabitants of our planet, have, in their fossil bones, left behind them imperishable monuments of the consummate skill with which they were constructed. Each limb and fragment of a limb form coordinate parts of a well-adjusted and perfect whole."

After reading these descriptions, it is not difficult to form a mental picture of the great beast laying siege to a tree, and to conceive the massive frame of the Megatherium convulsed with the mighty wrestling, every vibrating fibre reacting upon its bony attachment with the force of a hundred giants; extraordinary must be the strength and proportions of the tree if, when rocked to and fro, to right and left, in such an embrace, it can long withstand the efforts of its assailant. It yields, the roots fly up, the earth is scattered wide upon the surrounding foliage, and the tree comes down with a thundering crash, cracking and snapping the great boughs like glass. Then the coveted food is within reach, and the giant reaps the reward of his Herculean labours.

Sir Woodbine Parish thought that the Megatherium fed on the Agave, or American aloe.

Another form of extinct sloth found in the same region is the

---

[1] *Journal of Researches.* [2] *Bridgewater Treatise.*

Mylodon. Though of smaller size, it was much bigger than any living sloth, and attained a length of eleven feet. It has the same general structure, but the head and jaws are somewhat different, and more like the recent forms. A nearly perfect and original skeleton of Mylodon gracilis has been set up beside its huge relative's cast in the same gallery at the Natural History Museum. The crowns of its molar teeth are flat instead of being ridged ; hence its name, which signified " mill-toothed."

Yet another was the Scelidotherium [1] with its long limbs. Darwin obtained an almost entire skeleton of one of these. It was as large as a polar bear. Speaking of his discovery, he says, " The beds containing the fossil skeletons consist of stratified gravel and reddish mud ; a proof that the elevation of the land has been inconsiderable since the great quadrupeds wandered over the surrounding plains, and the external features of the country were then very nearly the same as now. The number of the remains of these quadrupeds embedded in the vast estuary deposits which form the Pampas and cover the granitic rocks of Banda Oriental must be extraordinarily great. I believe a straight line drawn in any direction through the country would cut through some skeleton or bones. As far as I am aware, not one of these animals perished, as was formerly supposed, in the marshes or muddy river-beds of the present land, but their bones have been exposed by the streams intersecting the subaqueous deposit in which they were originally embedded. We may conclude that the whole area of the Pampas is one wide sepulchre of these extinct gigantic quadrupeds." [2]

The genus Scelidotherium comprises a number of species and presents characters more or less intermediate between Megatherium and some other genera. The skull is low and elongated, and shows an approach to that of the modern ant-eater. The feet also are different from those of Megatherium (see Fig. 50).

These monster sloths inhabited South America during the

---

[1] Greek—*scelis*, limb ; *therion*, beast.　　　[2] *Journal of Researches.*

latest geological period, known as the Pleistocene. During part
of that time North America, as well as Northern Europe and
Asia, were invaded by a great ice-sheet, and an arctic climate
prevailed. It is therefore very probable that while the mammoth
and the mastodon were roaming over North America, giant sloths
and armadillos were monarchs of the southern continent. What
cause, or causes, led to the extermination of the giant sloths and
armadillos is still a matter of speculation. One writer suggests
an explanation that seems to deserve consideration. The
southern parts of this great continent are even now subject to
long-continued droughts, sometimes lasting for three years in

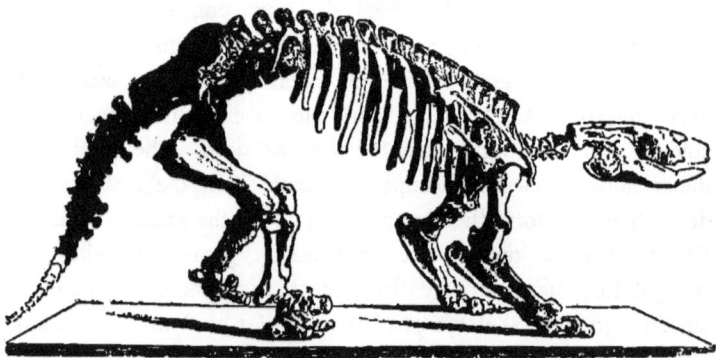

FIG. 50.—Skeleton of *Scelidotherium.* (After Capellini.)

succession, and bringing great destruction to cattle. In fact,
the discoveries related above were rendered possible by several
successive dry seasons. It is argued that the upright position
of most of the skeletons found *in situ* seems to suggest that
the creatures must have been mired in adhesive mud sufficiently
firm to uphold the ponderous bones after the flesh had decayed.
A long drought would bring the creatures from the drained
and parched country to the rivers, reduced by want of rain to
slender streams running between extensive mud-banks; and it
is possible that, in their anxious efforts to reach the water,
they may have only sunk deeper and deeper in the mud until

they were engulfed. This idea is strengthened by information supplied to Mr. Darwin when in these parts (recorded in his *Journal*). An eye-witness told him that during the *gran seco*, or great drought, the cattle in herds of thousands rushed into the Parana, and, being exhausted by hunger and thirst, were unable to crawl up the muddy banks, and so were drowned.

In the last great drought, from 1830 to 1832, it is probable (according to calculations made) that the number of animals that died was over one million and a half. The borders of all the lakes and streamlets in the province were long afterwards white with their bones.

In the year 1882 reports were published of the discovery of large footprints—supposed to be human—in a certain sandstone near Carson, Nevada, U.S. The locality was the yard of the State prison, and the tracks were uncovered in quarrying stone for building purposes. Many different kinds of tracks were found, some of which were made by an animal allied to the elephant; some resembled those of the horse and deer; others seem to have been made by a wolf, and yet others by large birds. Those supposed to have been made by human giants were in six series, each with alternate right and left tracks. The stride is from two and a half to over three feet, and each footprint is about eighteen inches long. Now, those who believed these tracks to be human must have found it hard to explain how a giant with a foot some eighteen inches long had a stride no longer than that of an ordinary man of to-day, to say nothing of the fact that the straddle was eighteen to nineteen inches! For these and other reasons Professor Marsh has exploded the idea of their having been made by men, and gave good reasons to show that they were probably made by a giant sloth, such as the Mylodon above mentioned, the remains of which have been discovered in the same strata. They agree in size, in stride, and in width between the right and left impressions, very closely with the tracks that a Mylodon would have made, and it seems that those of the fore feet were afterwards impressed by the hind feet, so that each track contains two impressions.

The reader who has some knowledge of natural history will not need to be told that the sloths of the present day, inhabiting the same region as their gigantic ancestors, are of small size, and live among the branches of the trees, together with the spider monkeys, howlers, and other apes. An interesting question to the evolutionist is—How did the change take place from the old huge and heavy types to the smaller and agile types of the present day? Can it be possible that the more difficult and tedious task of pulling down branches and even stems of trees, in order to devour the leaves, was abandoned for the simpler method of climbing up and feeding among the branches? It certainly looks as if a change of this kind had been instituted at some distant period in the past—distant, that is, to *us*, but not very remote geologically. The present method seems so much simpler that we need not be surprised at its adoption, for Nature is ever ready to encourage and assist those among the children of Life which can hit upon and adopt new and improved methods, either in obtaining food or repelling enemies, or other duties imposed upon them. Now, suppose that, in accordance with the well-known fact that variations in the offspring of animals are constantly cropping up, some considerably smaller variety of Megatherium, or Mylodon, or other now extinct type, appeared on the scene, and, by virtue of its comparative agility, could climb a tree and feed among the branches instead of pulling them down : then, as Darwin has so well explained, Nature would seize upon this accidental variation, and give it an advantage over its more awkward relations. Its offspring, too, would inherit the same characteristics, they would adopt the same habits, and, in time, as " natural selection " further increased these characters, by weeding out those that were unfit while fostering all those that were neither large nor clumsy in climbing trees, a new race of sloths would arise. This new race, it can well be imagined, would in time outstrip the old race in numbers, for successful races multiply while unsuccessful ones diminish. Victory is not always to the great and the strong, for cunning and quickness are often of more service than mere brute strength; and perhaps the sloths, as we now see them in the Brazilian forests, have hit upon "a new and original plan" by means of which the old colossal forms described above have been driven out of the field, and so exterminated by a process of competition. Such an explanation would be in thorough harmony with modern teaching, and, as the other suggestion about long-continued droughts, given on p. 184, may not appear satisfactory to some of our readers, we offer this theory for what it may be worth.

A few words about these modern sloths may not be out of place; for we shall better understand how they have succeeded in the struggle for existence when we know something of their manner of life; and in some ways they still resemble their great ancestors.

There are few animals which exhibit in a greater degree what appears to the careless observer to be *deformity* than the sloth, and none that have, on this account, been more maligned by naturalists. Buffon, and many of the older zoologists, were eloquent upon the supposed defects of the unfortunate sloth. These writers gravely asserted that when the sloth ascends a tree, for the purpose of feeding upon its leaves, it is so lazy that it will not quit its station until every trace of verdure is devoured. Some of them even went so far as to assert that when the sloth was compelled, after thus stripping a tree, to look out for a fresh supply of food, it would not take the trouble to descend the tree, but just allowed itself to drop from a branch to the ground. Even Cuvier, who ought to have known better, echoes this tale, and insinuates that Nature, becoming weary of perfection, "wished to amuse herself by producing something imperfect and grotesque," when the sloths were formed; and he proceeds, with great gravity, to show the "inconvenience of organisation," which, in his opinion, rendered the sloths unfit for the enjoyment of life.

It is quite true that, on the ground, these animals are about the most awkward creatures that can well be imagined. Their fore legs are much longer than their hind ones; all their toes are terminated by very long curved claws, and the general structure of the animal is such as to prevent them from walking in the manner of an ordinary quadruped, for they are compelled to rest on the sides of their hands and feet. Thus they appear the most helpless of animals, and their only means of progression consists in hooking their claws to some inequality in the ground, and thus dragging their bodies painfully along. But in their natural home, amongst the branches of trees, all these seeming disadvantages vanish—nay, the very peculiarities of structure which render the sloths objects of pity on the ground, are found to render them admirably adapted to their peculiar mode of life. The sloth is a small animal, rarely more than two feet in length, and covered with woolly hair—probably a protection against snakes, its only enemies. It spends nearly the whole of its life in the trees. There, safe from the prowling animals on the ground below, it hangs like a hammock from the bough, and even travels along the branches with its body downwards, using its long claws like grappling-irons.

It looks slothful enough when asleep, for then it resembles a bunch of rough hair, and a jumble of limbs close together, hanging to a branch; but when awake it is industrious in its search for nice twigs and leaves, and moves along with considerable activity. When the atmosphere is still, the sloth keeps to its tree, feeding on the leaves and twigs, but when there is wind, and the branches of neighbouring trees come in contact, the opportunity is seized, and the animal moves along the forest under the shady cover of the boughs. The Indians have a saying that " when the wind blows the sloth begins to crawl ; " and the reason is quite evident, for they cannot jump, but can hang, swing, and crawl suspended.

We now pass on to the old gigantic representative of the

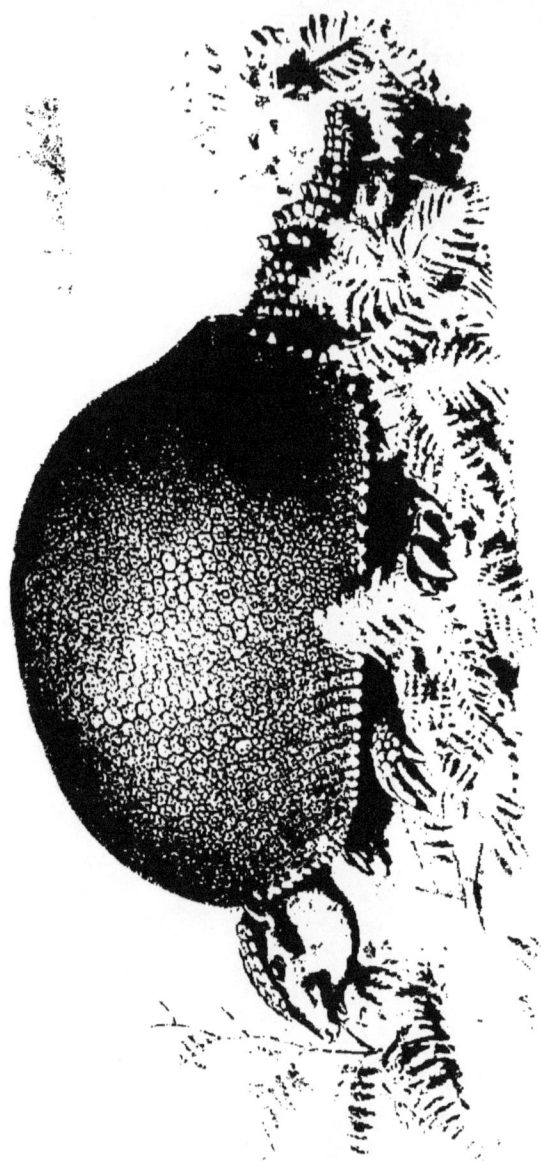

armadillo, the Glyptodon.[1]   To the eye it resembles more or less
an armadillo, and has a huge cuirass, or large plate of armour,
covering the whole of the body, but allowing the head to show in
front, while the legs come out beneath.   Both head and tail were
also protected with armour.   The great shield, or carapace, in
most of the extinct armadillos, is composed of long plates of
regular shape, closely united at their edges (sutures) so as to form
a solid piece.   It is evident, therefore, that this creature, having
no movable bands, as living armadillos have, could not roll itself
up into a ball.   The fore feet have thick, short toes, instead of
long ones, such as their modern representatives have; and
from this we may infer that they were not in the habit of burrowing or of seeking their food underground.   The family of
Glyptodonts seem to have been chiefly confined to the continent
of South America, but some species are known to have extended
their range as far as Mexico, and Texas into North America.   A
good deal of confusion has arisen with regard to the classification
of these old-fashioned armadillos, on account of the fact that
isolated specimens of their tails have often been found, and these
cannot always be referred to the right carapaces.   For example,
it should be pointed out here that the tail represented in Fig. 51
really belongs to another genus, known as Hoplophorus.[2]

In Glyptodon asper (Plate XIX.), the scutes of the carapace
had a beautiful rosette-like sculpture, while the sheath of the tail
was entirely composed of a series of movable rings, ornamented
with large projecting tubercles.   The vertebræ of the backbone
are almost entirely fused together into a long tube, and also are
joined to the under surface of the great shield, to which the ribs
are united.   The cheek-teeth are sixteen in number, four above
and four below on each side.   These are channelled with two
broad and deep grooves, which divide the surface into three
distinct lobes.   Hence the name of the animal.

[1] So named by Sir R. Owen, in reference to the sculptured aspect of the
grinding surface of the teeth.   Greek—*glupho*, I carve ; *odous, odontos*, tooth.

[2] Greek—*Hoplon*, armour ; *phero*, I bear.

The tessellated carapace of the Glyptodon was at first thought
to belong to the Megatherium, with which the remains were
associated, but Professor Owen clearly demonstrated the impossibility of this idea.

Fig. 51 represents Glyptodon clavipes (Owen) from the Pleistocene deposits of Buenos Ayres; but the reader will gain a much
better idea of the animal by inspecting the splendid specimen of
Glyptodon asper in the Natural History Museum, near the centre
window at the east end of the Pavilion (Glass-case Q on plan).

Plate XIX. is a restoration of another species by our artist.[1]

In the Museum of the Royal College of Surgeons (which the
reader is recommended to visit) there are several most valuable
specimens of these extinct armadillos from South America.

Armadillos belong, with sloths and ant-eaters, to the same

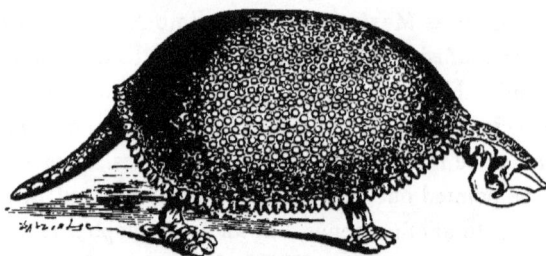

FIG. 51.—Extinct Gigantic Armadillo, *Glyptodon clavipes*, from Pleistocene
deposits, Buenos Ayres. (The tail sheath here represented probably
belongs to another genus, Hoplophorus.)

family of so-called toothless animals (edentata) with no front
teeth, though one or two forms really are toothless. Those of the
present day have their bony armour divided up into a series of
bands, so that they can roll themselves up, more or less, into
balls. They burrow under the ground, where they get their food
to a certain extent, and live a safe life, protected by their casque

[1] This plate is based on a beautiful drawing in a Spanish work, *Anales del
Museo publico Buenos Aires*. G. Burmeister, M.D., Phil. D. Tomo Segundo.

of mail. Their only enemies seem to be the monkeys, and one of the tricks of the young monkeys in the American forests is, when they find an armadillo away from home, to pull its tail unmercifully, and try to drag it about. Snakes cannot hurt them. Mr. Hudson, in his most interesting book, *A Naturalist in La Plata*, narrates how he watched an armadillo kill a snake and then devour it.

If we examine the anatomy of the armadillo, we shall find that its bones greatly resemble those of the sloth, but still there are a few differences. It is a burrowing animal, and therefore requires great power of scratching and tearing the ground. Why the colossal forms of armadillo should have become extinct and only small ones survived to the present time, is one of the many and perplexing problems presented by the study of extinct animals. One would have thought from its size and strength that the Glyptodon had been built, like Rome, for eternity.

CHAPTER XIII.

THE MAMMOTH.

> "Yes, where the huntsman winds his matin horn,
>   And the couched hare beneath the covert trembles ;
> Where shepherds tend their flocks, and grow their corn
>   Where fashion in our gay Parade assembles—
> Wild horses, deer, and elephants have strayed,
>   Treading beneath their feet old Ocean's races."
>
> HORACE SMITH.

MANY are the traditions and tales that have clustered round the Mammoth.[1] He is, however, no fabulous product of the imagination, like the dragon, for he has actually been seen in the flesh, and not only *seen*, but eaten, both by men and animals ! But, for all that, men's minds have been busy for centuries past making up tales, often of the wildest description, about him ; and it is little wonder that a creature whose bones are found in the soils and gravels, etc., over more than half the world, and whose body has been seen frozen in Siberian ice, should have given rise to many tales and superstitions. To students of folk-lore these legends are of considerable interest, and to some extent also to

---

[1] The word *Mammoth* is thought by Pallas and Nordenksiöld to be of Tartar origin. The former asserts that the name originated in the word *mamma*, which signifies earth (the Mammoth being found frozen in the earth). It was introduced into the languages of Western Europe about two centuries ago, from the Russian. But other writers have attempted to prove that it is a corruption of the Arabic word *Behemoth*, or "great beast," which in the Book of Job signifies an unknown animal. In an ancient Chinese work, of the fifth century before Christ, it is spoken of under the name *tien-schu*, that is to say, "the mouse which hides itself." The Chinese legends are referred to on p. 199.

men of science. To the latter, however, one of its many points of interest is that palæontology may be said to have been founded on the Mammoth. Cuvier, the illustrious founder of the science of organic remains, was enabled, by his accurate and minute knowledge of the structures of living animals, to prove to his astonished contemporaries that the Mammoth bones and teeth, so plentifully discovered in Europe, were not such as could have belonged to any living elephant, and consequently that there must have existed, at some previous period in the world's history, an elephant of a different kind, and quite unknown to naturalists. This was a new idea, and accordingly one that met with opposition as well as incredulity.

It was thought in those days that whatever animals lived in the past *must* have resembled those now inhabiting the world, and the idea of extinct types unknown to man, and unknown to the regions where their bones were found embedded below the soil, was of so novel and startling a character as to appear incredible. Besides, the Mosaic account of Creation made no direct reference to extinct animals, and therefore the notion was not to be entertained.

It is amusing to note the devices to which people resorted in order to combat this revolutionary teaching. Thus, when Cuvier first announced the discovery of the fossil remains of the elephant, hippopotamus, and rhinoceros in the superficial deposits of continental Europe, he was gravely reminded of the elephants introduced into Italy by Pyrrhus in the Roman wars, and afterwards in the Roman triumphal processions or the games at the Colosseum.

It was only by means of minute anatomical differences that he was able to show that the bones and teeth of these elephants must have belonged to a species unlike those now living. But these differences proved too subtle for even scientific men to appreciate, so slight was their knowledge of anatomy compared with his; so that they were either disallowed or explained away.

But he was not to be beaten, and appealed to the fact that

similar remains occurred in Great Britain, whither neither Romans nor others could have introduced such animals. These are his words: "If, passing across the German Ocean, we transport ourselves into Britain, which in ancient history by its position could not have received many living elephants besides that one which Cæsar brought thither, according to Polycenus; we shall, nevertheless, find these fossils in as great abundance as on the Continent."

Another crushing answer to the absurd explanations of Cuvier's countrymen was added by the sagacious Dean Buckland, who pointed out that in England, as on the Continent, the remains of elephants are accompanied by the bones of the rhinoceros and hippopotamus, animals which not even Roman armies could have subdued or tamed! Owen also adds that the bones of fossil elephants are found in Ireland, where Cæsar's army never set foot.

It was in 1796 that Cuvier announced that the teeth and bones of the European fossil elephants were distinct in species from both the African and the Indian elephant, the only two living species (El. africanus and El. indicus). This fundamental fact opened out to him new views about the creation of the world and its inhabitants, and a rapid glance over other fossil bones in his collection showed him the truth and the value of this great idea (namely, the existence of extinct types), to which he consecrated the rest of his life. Thus palæontology may be said to have been founded on the Mammoth.

The fossil remains of elephants have, on account of their common occurrence in various parts of the world, attracted a great deal of attention, both from the learned and the unlearned. In the North of Europe they have been found in Ireland, in Germany; in Central Europe, in Poland, Middle and South Russia, Greece, Spain, Italy; also in Africa, and over a large part of Asia. In the New World they have been found abundantly in North America. But in the frozen regions of Siberia its tusks, teeth, and bones are met with in very great abundance. According to Pallas, the great Russian savant, there is not in the whole of Asiatic Russia, from the Don to the extremity of the Tchutchian promontory, any brook or river on the banks of which some bones of elephants and other animals foreign to these regions have not been found. The primæval elephants (Mammoth, Mastodon, etc.) appear to have formerly ranged over the whole northern hemisphere of the globe, from the fortieth parallel to the sixtieth, and possibly to near the seventieth degree of latitude.

Just as the North American Indian regards the great bones of Professor Marsh's extinct Eocene mammals that peep out from the sides of buttes and cañons, as belonging to his ancestors, so we find that in all parts of the world the bones of extinct elephants have, on account of their great size (and partly from a certain resemblance, in some, to bones of the human skeleton), been regarded as testifying to the former existence of giants, heroes, and demigods. To the present day the Hindoos consider such remains as belonging to the *Rakshas*, or Titans,—beings that figure largely in their ancient writings. Theophrastus, of Lesbos, a pupil of Aristotle, appears to have been the first to record the discovery of fossil ivory and bones. These were probably obtained by the country people from certain deposits in the neighbourhood, and are mentioned five hundred years later by Pausanias. Several Greek legends and traditions appear to be founded on such discoveries.

Thus the Greeks mistook the knee-bone of an elephant for that of Ajax. In like manner the supposed body of Orestes, thirteen feet in length, discovered by the Spartans at Tegea, doubtless was the skeleton of some elephant. In the isle of Rhodes, in Sicily, and near Palmero, Syracuse, and at many other places, similar remains have afforded a basis for stories of giants. In fact, so much has been said by old writers on this subject, that whole volumes might be filled with such matter. Let one or two examples suffice.

In the year 1613 some workmen in a sand-pit near the castle
of Chaumont, not far from St. Antoine, found some bones
(probably of the Mammoth or Mastodon) of the nature of which
they were entirely ignorant, and many of them they broke up.
But a certain surgeon named Mazuyer, hearing of the discovery,
bought the bones, and announced that he had himself discovered
them in a tomb thirty feet long, bearing in Gothic characters
the inscription, "Teutobochus Rex." This was a barbarian king
who invaded Gaul at the head of the Cimbri, and was defeated
near Aix, in Provence, by Marius, who brought him to Rome to
grace his triumphal procession. Mazuyer reminded his credulous
readers that, according to the testimony of Roman authors, the
head of this king was larger than any of the trophies borne upon
the lances in triumph, and for a time his marvellous story was
accepted. The skeleton of this pretended giant-king was exhibited in many cities of France and Germany, and also before
Louis XII., who took great interest in it. The imposture was
detected and exposed by Riolan, and thus a great controversy
arose, and numerous pamphlets were written on both sides. The
skeleton remained at Bordeaux till the year 1832, when it was
sent to the Museum of Natural History at Paris, where it may
still be seen. It is needless to say that, on its arrival there,
M. Blainville at once recognised it as being that of an elephant
—a Mastodon, in fact.

Another giant-story may be narrated as follows. In the year
1577 some large bones were discovered, through the uprooting of
an oak by a storm, in the Canton of Lucerne, in Switzerland.
These bones were afterwards declared by the celebrated physician
and professor at Basle, Felix Plater, to be those of a giant. This
learned man estimated the height of the giant at nineteen feet !
and made a drawing thereof, which he sent to Lucerne. The
bones have since nearly all vanished, but Blumenbach, at the
beginning of this century, saw enough of them to prove their
elephantine nature. The good people of Lucerne, however,

being reluctant to abandon their giant, have, since the sixteenth century, made him the supporter of their city arms.

The Church of St. Christopher, at Valence, possessed an elephant's tooth, which was shown as the tooth of St. Christopher. As this relic was "bigger than a man's fist," it is difficult to picture what idea the people entertained of their saint!

In 1564 two peasants observed on the banks of the Rhone, along a slope, some great bones sticking out of the ground. These they carried to the neighbouring village, where they were examined by Cassanion, who lived at Valence, and was the author of a treatise on giants (*De Gigantibus*). Cuvier concluded from this writer's description of the tooth that it belonged to an elephant.

Otto de Guericke, famous as the inventor of the air-pump, in 1663 witnessed the discovery of a fossil elephant, with its tusks preserved. These he mistook for horns; so did even the illustrious Leibnitz, who created out of his own imagination a strange animal, with a great horn in the middle of its forehead, as the creature to which these remains belonged! One is reminded of Bret Harte's amusing *jeu d'esprit, The Society upon the Stanislaus*—

> " Then Brown he read a paper, and he reconstructed there,
> From those same bones, an animal that was extremely rare ; "

and how the members of this learned society came to blows over their fossil bones, and hurled them at one another—"till the skull of an old mammoth caved the head of Thomson in." But in this case, the "animal that was extremely rare" was believed in for a long time, and Leibnitz's "fossil unicorn" was universally accepted throughout Germany for more than thirty years. At last, however, a complete skeleton of a Mammoth was discovered, and recognised as belonging to an elephant ; but the unicorn was not given up without a keen controversy.[1]

---

[1] The writer is indebted for much of the information here given with regard to the discoveries of Mammoth bones, and legends founded thereon, to M. Figuier's *World before the Deluge*.

Near the city of Constadt, in the year 1700, a great quantity of
bones and tusks of elephants were discovered, after excavations
had been made by order of the reigning duke, who had been
informed by a soldier of Würtemberg of the presence of bones in
the soil.   In this way some sixty tusks were unearthed.   The
whole ones were preserved, but those which were broken were
given to the Court physician, who made use of them for medicinal
purposes.   After this the " Ebur fossile," or " Unicornu fossile,"
was freely used by the German doctors, until the discovery of the
bone-caves of the Hartz, when it became too abundant to pass
for true unicorn, and consequently lost much of its repute.

In our own country elephantine remains have also given rise to
strange tales.   The village of Walton, near Harwich, is famous
for the abundance of Mammoth remains, which lie along the base
of the sea-cliffs, mixed with the bones of horses, oxen, and deer.
" The more bulky of these fossils," says Professor Owen, "appear
to have early attracted the notice of the curious.   Lambard, in
his *Dictionary*, says that 'in Queen Elizabeth's time bones were
found, at Walton, of a man whose skull would contain five pecks,
and one of his teeth as big as a man's fist, and weighed ten
ounces.   These bones had sometimes bodies, not of beasts, but
of men, for the difference is manifest.'".

According to the same authority, there is reason to believe
that instances have occurred in Great Britain in which, with due
care and attention, a more or less entire skeleton of the Mammoth
might have been secured.   He mentions the case of the discovery
of a number of Mammoth bones by some workmen in a brick-
ground, near the village of Grays, in Essex.   But most unfortunately, in their ignorance, they broke up these valuable relics,
and sold the fragments, for three half-pence a pound, to a dealer
in old bones !   This somewhat lucrative traffic went on for over
half a year before the matter came to the notice of Mr. R. Ball,
F.G.S., who recovered some fine bones from the men, and thus
rescued them from the destruction that awaited them.

It is greatly to be hoped that some day our National Treasure House at South Kensington may be enriched with a complete Mammoth skeleton from British soil.

The Chinese, as might be expected, heard of the Mammoth long before Europeans did, and they have some strange stories about it. In the northern part of Siberia, so great is the abundance of Mammoth tusks, that for a very long period there has been a regular export of Mammoth ivory, both eastward to China and westward to Europe. Even in the middle of the tenth century an active trade was carried on at Khiva in fossil ivory, which was fashioned into combs, vases, and other objects, as related by an Arab writer of that time. Middendorf reckoned that the number of fossil tusks which have yearly come into the market, during the last two centuries, has been at least a hundred pairs—an estimate which Nordenskiöld considers as well within the mark. They are found all along the line of the shore between the mouth of the Obi and Behring Straits, and the further north a traveller goes, the more numerous does he find them. The soil of Bear Island and of the Liachoff Islands (New Siberia) is said to consist only of sand and ice with such quantities of Mammoth bones that it appears as if they were almost made up of bones and tusks. Every summer numbers of fishermen make for these islands to collect fossil ivory, and during the winter immense caravans return laden with Mammoth tusks. The convoys are drawn by dogs, and in this way the ivory reaches both the ancient Eastern and the newer Western markets.

It is evident from the Chinese legends that the frozen bodies of Mammoths have for ages past been either seen by, or reported to, members of the celestial empire, for it is mentioned in some of their old books as an animal that lives underground. In a great Chinese work on natural history, written in the sixteenth century, the following quaint description occurs : "The animal named *tien-schu*, of which we have already spoken, in the ancient work upon the ceremonial entitled *Lyki* [a work of the fifth century before Christ] is called also *fyn-schu*, or *yn-schu*, that is to say, 'the mouse that hides itself.' It always lives in subterranean caverns; it resembles a mouse, but is of the size of a buffalo or ox. It has no tail; its colour is dark; it is very strong, and excavates caverns in places full of rocks and forests." Another writer says, "The *fyn-schu* haunts obscure and unfrequented places. It dies as soon as it is exposed to the rays of the sun or moon; its feet are short in proportion to its size, which causes it to walk badly. Its tail is a Chinese ell in length. Its eyes are small, and its neck short. It is very stupid and sluggish. When the inundations of the river *Tamschuann-tuy* took place [in 1571] a great many *fyn-schu* appeared in the plain; it fed on the roots of the plant *fu-kia*."

An old Russian traveller, who, in 1692, was sent by Peter the Great as ambassador to the Emperor of China, mentions the discovery of the heads and legs of Mammoths in frozen soil. After referring to these discoveries, he says, "Concerning this animal there are very different reports. The heathens of Jakutsk, Tungus, and Ostiaks say that they continually, or at least, by reason of the very hard frosts, mostly live underground, where they go backwards and forwards; to confirm which they tell us that they have often seen the earth heaved up when one of these beasts was upon the march, and, after he passed, the place sink in, and thereby make a deep pit. They further believe that if this animal comes so near to the surface of the frozen earth as to . smell the air, he immediately dies, which they say is the reason that several of them are found dead on the high banks of the river, where they unawares came out of the ground. This is the opinion of the infidels concerning these beasts, which are never seen. But the old Siberian Russians affirm that the Mammoth is very like the elephant, with this difference only, that the teeth of the former are firmer, and not so straight as those of the latter. . . . By all I could gather from the heathens, no person ever saw one of these beasts alive, or can give any account of its shape; so that all we heard said on this subject arises from bare conjecture only."

But making all allowance for the gross absurdities of these accounts, it is clear that they are based on descriptions—probably by the Tungusian fishermen—of carcases that have been washed out of the frozen soil by rivers in flood time. Now that we are in possession of trustworthy accounts, we can understand how these strange tales arose among an ignorant and superstitious people, such as the fishermen of these inhospitable shores.

We will now put before the reader the true accounts given by Adams [1] and Benkendorf.

In 1799 a Tungusian, named Schumachoff, who generally went to hunt and fish at the peninsula of Tamut after the fishing season of the Lena was over, had constructed for his wife some cabins on the banks of the lake Oncoul, and had embarked to seek along the coasts for Mammoth tusks. One day he saw among the blocks of ice a shapeless mass, but did not then discover what it was. In 1800 he perceived that this object was more disengaged from the ice, and that it had two projecting parts; and towards the end of the summer of 1801 the entire side of the animal and one of his tusks were quite free from ice. In 1803 the enormous mass fell by its own weight on a bank of sand. It was a frozen Mammoth! In 1804 Schumachoff came to his Mammoth, and having cut off the tusks, exchanged them with a merchant for goods. Two years afterwards Mr. Adams, the narrator of the story, traversed these distant and desert regions, and found the Mammoth still in the same place, but sadly mutilated. The people of the neighbourhood had cut off the flesh, and fed their dogs with it during the scarcity. Wild beasts, such as white bears, wolves, and foxes, also had fed on it, and the traces of their footsteps were seen around. The skeleton was complete

[1] Abridged from *Memoirs of the Imperial Academy of Sciences of St. Petersburg*, vol. v. London, 1819.

all except one leg, but the flesh had almost all gone.   The head
was covered with a dry skin, one of the ears was seen to be
covered with a tuft of hairs.   All these parts suffered more or
less injury in transport for a distance of 7330 miles to St.
Petersburg, yet the eyes have been preserved.   This Mammoth
was a male, with a long mane on its neck, but both tail and
proboscis had disappeared.   The skin is of a dark grey colour,
covered with a reddish wool and black hairs.   The entire carcase
was nine feet four inches high.   The skin of the side on which
the carcase had lain was detached by Mr. Adams, for it was well
preserved, but so heavy was it that ten persons found great
difficulty in transporting it to the shore.   The white bears, while
devouring the flesh, had trodden into the ground much of the
hair belonging to the carcase, but Mr. Adams was able by digging
to procure about sixty pounds' weight of hair.   In a few days the
work was completed, and he found himself in possession of
a treasure which amply compensated him for the fatigues and
dangers of the journey as well as the expense of the enterprise.
When first seen, this Mammoth was embedded in clear pure ice,
which forms in that coast escarpments of considerable thickness,
sloping towards the sea, the top of which is covered with moss
and earth.   If the account of the Tungusians can be trusted, the
carcase was some way below the surface of the ice when first
seen.   Arrived at Takutsk, Mr. Adams purchased a pair of tusks
which he believed to belong to this Mammoth, but there is reason
to doubt whether he did get the right tusks.   They are nine feet
six inches long.

The skeleton of this specimen, the fame of which may be said
to have spread all over the world, is now set up in the Museum
of the St. Petersburg Academy, and the skin still remains attached
to the head and feet.   A part of the skin and some of the hair
were sent by Mr. Adams to Sir Joseph Banks, who presented them
to the Museum of the Royal College of Surgeons.[1]   A photograph

---

[1] A specimen of the hair of a mammoth may be also seen at the Natural

of the skeleton as it now stands, may be seen on the wall of the
big Geological Gallery at South Kensington (No. I. on plan), near
the specimens of Mammoth tusks.   But it should be pointed out
that *the tusks are put on the wrong way ;* for they curve outwards
instead of inwards, thus presenting a somewhat grotesque appearance.   For this reason we have not reproduced the familiar
woodcut based on an engraving in the memoir already referred

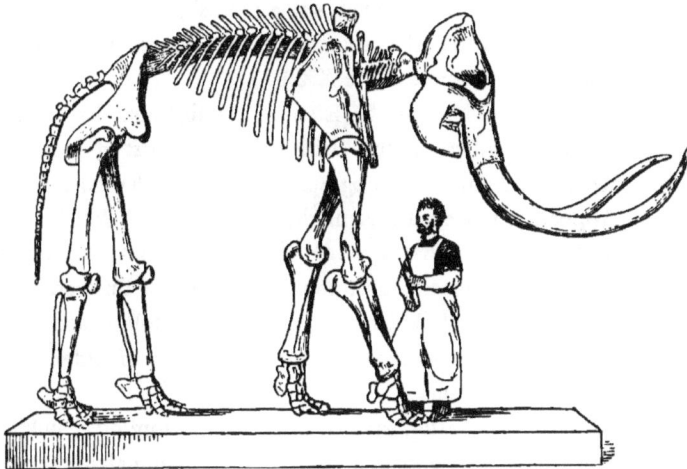

FIG. 52.—Skeleton of Mammoth, *Elephas primigenius* (partly restored), in
the Museum at Brussels.   Drawn from a photograph, by J. Smit.

to.[1]   But we give, instead, a sketch taken from a photograph (also
on the wall in gallery No. I.) of a fine skeleton in the Brussels
Museum (Fig. 52).   Here the tusks are seen correctly inserted.
We must also draw the reader's attention to the remarkably fine

History Museum (pier case 31) in a tall glass jar. It came from frozen soil,
Behring Strait. By the side of this will be seen, in a glass box, a portion of
the skin of a mammoth, from the banks of the river Alaseja, Province of
Yakutsk, Siberia. It exhibits the under fur, the long hair having entirely
disappeared.

[1] Fig. 32 in Part I. of the *Guide to the Exhibition Galleries in the Department of Geology and Palæontology in the British Museum (Natural History),
Cromwell Road.* (Price 1s.) This most useful guide should be consulted by
the reader.

specimen (glazed case E on plan) consisting of the skull and both tusks complete, found at Ilford in Essex.

Adams's specimen was, Dr. Woodward thinks, an old individual, and its tusks had curved upwards so much as to be of little use. In younger ones they were less curved. The hair that still remains on the skin of the St. Petersburg specimen is of the colour of the camel, very thick-set and curled in locks. Bristles of a dark colour are interspersed, some reddish, and some nearly black. The colour of the skin is a dull black, as in living elephants (see restoration, Plate XX.).

Remains of the Mammoth (Elephas primigenius) have been found in great numbers in the British Isles. A list of localities (from Mr. Leith Adams's monograph on fossil elephants) is given in the Appendix, but even this might be extended. Mr. Samuel Woodward calculated that upward of two thousand grinders of elephants have been dredged up during a period of thirteen years upon the oyster-bed off Hasborough, on the Norfolk coast. But many of these doubtless belong to other species of older date, such as Elephas antiquus.

Dr. Bree, of Colchester, says that the sea-bottom off Dunkirk, whence he has made a collection, is so full of mammalian remains that the sailors speak of it as "the Burying-ground."

The remains of the Mammoth occur over a very large geographical area—fully half the globe.

By far the most important discovery of a frozen Mammoth is that of a young Russian engineer, Benkendorf by name, who was an eye-witness of its resurrection, though, most unfortunately, he was unable either to procure his specimen, as Mr. Adams did, or to make drawings of it. Being employed by the Russian Government in making a survey of the coast off the mouth of the Lena and Indigirka rivers, he was despatched up the latter river in 1846, in command of a small steam-cutter. The following is a translation of the account which he wrote to a friend in Germany.

" In 1846 there was unusually warm weather in the north of Siberia. Already in May unusual rains poured over the moors and bogs, storms shook the earth, and the streams carried not only ice to the sea, but also large tracts of land, thawed by the masses of warm water fed by the southern rains. . . . We steamed on the first favourable day up the Indigirka ; but there were no thoughts of land ; we saw around us only a sea of dirty brown water, and knew the river only by the rushing and roaring of the stream. The river rolled against us trees, moss, and large masses of peat, so that it was only with great trouble and danger that we could proceed. At the end of the second day, we were only about forty versts up the stream ; some one had to stand with the sounding-rod in hand continually, and the boat received so many shocks that it shuddered to the keel. A wooden vessel would have been smashed. Around us we saw nothing but the flooded land for eight days. We met with the like hindrances until at last we reached the place where our Jakuti were to have met us. Further up was a place called Ujandina, whence the people were to have come to us ; but they were not there, prevented evidently by the floods.

" As we had been there in former years, we knew the place. But how it had changed ! The Indigirka, here about three versts wide, had torn up the land and worn itself a fresh channel ; and when the waters sank we saw, to our astonishment, that the old river-bed had become merely that of an insignificant stream. This allowed me to cut through the soft earth, and we went reconnoitring up the new stream, which had worn its way westwards. Afterwards we landed on the new shore, and surveyed the undermining and destructive operation of the wild waters, that carried away, with extraordinary rapidity, masses of soft peat and loam. It was then that we made a wonderful discovery. The land on which we were treading was moorland, covered thickly with young plants. Many lovely flowers rejoiced the eye in the warm beams of the sun, that shone for twenty-two out of the twenty-four hours. The stream rolled over and tore up the soft wet ground like chaff, so that it was dangerous to go near the brink. While we were all quiet, we suddenly heard under our feet a sudden gurgling and stirring, which betrayed the working of the disturbed waters. Suddenly our jäger, ever on the outlook, called loudly, and pointed to a singular and unshapely object, which rose and sank through the disturbed waters. I had already remarked it, but not given it any attention, considering it only driftwood. Now we all hastened to the spot on the shore, had the boat drawn near, and waited until the mysterious thing should again show itself. Our patience was tried, but at last a black, horrible, giant-like mass was thrust out of the water, and we beheld a colossal elephant's head, armed with mighty tusks, with its long trunk moving in the water in an unearthly manner, as though seeking for something lost therein. Breathless with astonishment, I beheld the monster hardly twelve feet from me, with his half-open eyes yet showing the whites. It was still in good preservation.

"'A mammoth! a mammoth!' broke out the Tschernomori; and I shouted, 'Here, quickly. Chains and ropes!' I will go over our preparations for securing the giant animal, whose body the water was trying to tear from us. As the animal again sank, we waited for an opportunity to throw the ropes over his neck. This was only accomplished after many efforts. For the rest we had no cause for anxiety, for after examining the ground I satisfied myself that the hind legs of the Mammoth still stuck in the earth, and that the waters would work for us to unloosen them. We therefore fastened a rope round his neck, threw a chain round his tusks, that were eight feet long, drove a stake into the ground about twenty feet from the shore, and made chain and rope fast to it. The day went by quicker than I thought for, but still the time seemed long before the animal was secured, as it was only after the lapse of twenty-four hours that the waters had loosened it. But the position of the animal

was interesting to me; it was standing in the earth, and not lying
on its side or back as a dead animal naturally would, indicating
by this the manner of its destruction. The soft peat or marsh
land, on which he stepped thousands of years ago, gave way
under the weight of the giant, and he sank as he stood on it,
feet foremost, incapable of saving himself; and a severe frost
came and turned him into ice, and the moor which had buried
him. The latter, however, grew and flourished, every summer
renewing itself. Possibly the neighbouring stream had heaped
over the dead body plants and sand. God only knows what
causes had worked for its preservation; now, however, the stream
had brought it once more to light of day, and I, an ephemera of
life compared with this primæval giant, was sent by Heaven just
at the right time to welcome him. You can imagine how I
jumped for joy.

"During our evening meal, our posts announced strangers—a
troop of Jakuti came on their fast, shaggy horses. They were our
appointed people, and were very joyful at the sight of us. Our
company was augmented by them to about fifty persons. On
showing them our wonderful capture, they hastened to the stream,
and it was amusing to hear how they chattered and talked over
the sight. The first day I left them in quiet possession, but when,
on the following, the ropes and chains gave a great jerk, a sign
that the Mammoth was quite freed from the earth, I commanded
them to use their utmost strength and bring the beast to land.
At length, after much hard work, in which the horses were
extremely useful, the animal was brought to land, and we were
able to roll the body about twelve feet from the shore. The
decomposing effect of the warm air filled us all with astonishment.

"Picture to yourself an elephant with a body covered with
thick fur, about thirteen feet in height, and fifteen in length, with
tusks eight feet long, thick, and curving outward at their ends,[1]
a stout trunk of six feet in length, colossal limbs of one and a

[1] This must be incorrect (see p. 203).

half feet in thickness, and a tail, naked up to the end, which was covered with thick tufty hair. The animal was fat and well-grown; death had overtaken him in the fulness of his powers. His parchment-like, large, naked ears, lay fearfully turned over the head; about the shoulders and the back he had stiff hair, about a foot in length, like a mane. The long outer hair was deep brown and coarsely rooted. The top of the head looked so wild, and so penetrated with pich[1] that it resembled the rind of an old oak tree. On the sides it was cleaner, and under the outer hair there appeared everywhere a wool, very soft, warm and thick, and of a fallow-brown colour. The giant was well protected against the cold. The whole appearance of the animal was fearfully strange and wild. It had not the shape of our present elephants. As compared with our Indian elephants, its head was rough, the brain-case low and narrow, but the trunk and mouth were much larger. The teeth were very powerful. Our elephant is an awkward animal, but compared with this Mammoth it is as an Arabian steed to a coarse, ugly dray-horse. I could not divest myself of a feeling of fear as I approached the head; the broken, widely-open eyes, gave the animal an appearance of life, as though it might move in a moment and destroy us with a roar. . . . The bad smell of the body warned us that it was time to save of it what we could, and the swelling flood, too, bid us hasten. First of all we cut off the tusks, and sent them to the cutter. Then the people tried to hew off the head, but notwithstanding their good will, this work was slow. As the belly of the animal was cut open the intestines rolled out, and then the smell was so dreadful that I could not overcome my nauseousness, and was obliged to turn away. But I had the stomach separated, and brought on one side. It was well filled, and the contents instructive and well preserved. The principal were young shoots of the fir and pine; a quantity of young fir-cones, also in a chewed state, were mixed with the mass. . . . As we were

---

[1] " Und mit Pech so durchgedrungen."

eviscerating the animal, I was as careless and forgetful as my Jakuti, who did not notice that the ground was sinking under their feet, until a fearful scream warned me of their misfortune, as I was still groping in the animal's stomach. Shocked, I sprang up, and beheld how the river was burying in its waves our five Jakuti and our laboriously saved beast. Fortunately, the boat was near, so that our poor workpeople were all saved, but the Mammoth was swallowed up by the waves, and never more made its appearance."

Much may be learned from this highly interesting account; it contains the key to several questions which otherwise might have remained unsolved. Let us see what conclusions can be derived therefrom. *First,* its position and perfect state of preservation are sufficient to prove that it was buried where it died. It sank in a marsh, probably during the summer. Then came the cold of winter ; the carcase, together with the ground around it, was frozen so that decomposition was arrested, and frozen it must have remained for many centuries till the day when M. Benkendorf came across it. Or it may have been buried up in a snow-drift which in time became ice.

In the region where frozen Mammoths occur (and there are at least nine cases on record), a considerable thickness of frozen soil may be found at all seasons of the year ; so that if a carcase be once embedded in mud or ice, its putrefaction may be arrested for indefinite ages. According to one authority, the ground is now permanently frozen even to the depth of four hundred feet at the town of Jakutsh, on the western bank of the river Lena. Throughout a large part of Siberia the boundary cliffs of the lakes and rivers consist of earthy materials and ice in horizontal layers. Middendorf bored to the depth of seventy feet, and after passing through much frozen soil mixed with ice, came down upon a solid mass of pure transparent ice, the depth of which he was unable to ascertain.

The year 1846, when M. Benkendorf saw his Mammoth, was

P

exceptional on account of its unusually warm summer, so that the ground of the tundra region thawed, and was converted into a morass. Had any Mammoths been alive then, and strayed beyond the limits of the woods into the tundra, probably some of them would have been likewise engulphed, and, when once covered up and protected from the decaying action of the air, the cold of the next winter would have frozen their carcases as this one must have been frozen up.

Truly, "there is nothing new under the sun," and the present highly useful method of freezing meat and bringing it over from America or New Zealand to add to our insufficient home supplies, is but a resort to a process employed by Nature long before the age of steamships, and perhaps even before the appearance of man on the earth !

Secondly, with regard to the food of the Mammoth, Benkendorf's discovery is of great service in solving the question how such a creature could have maintained its existence in so inhospitable and unpromising a country. The presence of fir-spikes in the stomach is sufficient to prove that it fed on vegetation such as is now found at the northern part of the woods as they join the low treeless tundra in which the body lay buried.

Before this discovery the food of the Mammoth was unknown, and all sorts of theories were devised in order to account for its remains being found so far north. Some thought that the Mammoth lived in temperate regions, and that the carcases were swept down by great floods into higher and colder latitudes. But it would be impossible for the bodies to be hurried along a devious course for so many miles without a good deal of injury, and probably they would fall to pieces on the way. But, as Professor Owen has so convincingly argued, there is no reason why herds of Mammoths should not have obtained a sufficient supply of food in a country like the southern part of Siberia, where trees abound in spite of the fact that during a great part of the year it is covered with snow. And this is his line of reasoning. The molar teeth of the elephant show a highly complicated and peculiar structure, and there are no other quadrupeds that feed to such an extent on the woody fibre of the branches of trees. Many mammals, as we know, eat the leaves of trees; some gnaw the bark; but elephants alone tear down and crunch the branches. One would think there was but little nourishment to be got from such. But the hard vertical plates of their huge grinders enable them to pound up the tough vegetable tissue and render it more or less palatable. Of course, the foliage is the most tempting, but where foliage is scarce something more is required.

Now, in the teeth of the Mammoth the same principle of construction is observed, only with greater complexity, for there are more of these grinding plates and a larger proportion of dense enamel. Hence the inference seems unmistakable that the extinct species fed more largely on woody fibre than does the elephant of to-day. Forests of hardy trees and shrubs still grow upon the frozen soil of Siberia, and skirt the banks of the Lena as far north as the sixtieth parallel of latitude.

If the Mammoth flourished in temperate latitudes only, as formerly suggested, then its thick shaggy coat becomes superfluous and meaningless; but if it lived in the region where its body has been found, then the argument from its teeth, and the fir-spikes found in its stomach, is confirmed by the nature of its skin, and all the old difficulties vanish. Professor Owen considers that we may safely infer that, if living at the present day, it would find a sufficient supply of food at all seasons of the year in the sixtieth parallel, and even higher. Perhaps they migrated north during the summer; and, judging from the present limits of arboreal vegetation, they may have been able to subsist even in latitude 70° north, for at the extreme points of Lapland pines attain a height of sixty feet.[1]

It is often no easy matter to form conclusions with regard to

[1] Sir Henry Howorth, in his *Mammoth and the Flood*, suggests another theory, and gives some valuable information.

the habits of extinct animals ; and too much reliance must not be placed on arguments derived from the habits of their living descendants or their near relations. The older geologists fell into this mistake with regard to the Mammoth, as did even Cuvier. Modern elephants are at present restricted to regions where trees flourish with perennial foliage, and, therefore, it was argued that there must have been a change of climate—either gradual or sudden, in the country of the Mammoth.

Cuvier, who believed in sudden revolutions on the earth's surface, argued that the Mammoth could not possibly have lived in Siberia as it is now ; and that, at the very moment when the beast was destroyed, the land was suddenly converted into a glacial region ! ("C'est donc le même instant qui a fait périr les animaux, et qui a rendu glacial le pays qu'ils habitaient, cet événement a été subit, instantané, sans aucune gradation."[1]) Sir Charles Lyell argued, from geological evidence with regard to the rise of land along the Siberian coast, that the climate had become somewhat more severe, and that finally the Mammoth, though protected by its shaggy coat, died out on account of scarcity of food.[2]

Professor Owen is unwilling to believe that such changes as these brought about the final extinction of the Mammoth, and he concludes that it was quite possible for such an animal to have flourished as near to the North Pole as is compatible with the growth of hardy trees or shrubs.

"The fact seems to have been generally overlooked, that an animal organised to gain its subsistence from the branches or woody fibre of trees, is thereby rendered independent of the seasons which regulate the development of leaves and fruit ; the forest food of such a species becomes as perennial as the lichens that flourish beneath the winter snows of Lapland ; and, were such a quadruped to be clothed, like the reindeer, with a natural

[1] *Ossemens Fossiles,* tom. i. p. 108.
[2] See *The Principles of Geology,* vol. i. chap. x.

garment capable of resisting the rigours of an arctic winter, its adaptation for such a climate would be complete. . : . The wonderful and unlooked-for discovery of an entire Mammoth, demonstrating the arctic character of its natural clothing, has, however, confirmed the deductions which might have been legitimately founded upon the localities of its most abundant remains, as well as upon the structure of its teeth, viz. that, like the Reindeer and Musk Ox of the present day, it was capable of existing in high northern latitudes." [1]

The problem of the extinction of the Mammoth is not an easy one to solve. We can hardly account for its disappearance by calling in geographical changes by which its range became restricted, and its food supply diminished, so that in the competition with other herbivorous animals this primæval giant "went to the wall," as the saying is. Nor does Lyell's appeal to a change in climate, by which the cold of Siberia became too intense even for the Mammoth, seem quite satisfactory, especially when we remember how very far north fir trees range (p. 211).

The Mammoth, probably, was endowed with a fairly tough constitution. In Siberia it fed on fir trees. In Kentucky it fared better, and was surrounded by such vegetation as now flourishes in that temperate region. In the valley of the Tiber (where also its remains are found), though during the "Glacial period" the temperature was, doubtless, lower than at present, we cannot imagine that an arctic climate prevailed. Thus we see that it was capable of flourishing in various and widely separated regions where the conditions of climate and food supply could hardly have been similar.

Professor Boyd Dawkins, whose views we are adopting here,[2] considers that the Mammoth was exterminated by man—a simple

[1] *A History of British Fossil Mammals and Birds*, by Richard Owen, F.R.S., etc. London, 1846.

[2] *Popular Science Review*, vol. vii. p. 275 (1868).

solution of the question, which seems to present no difficulties. That it was hunted by the primitive folk of the "Reindeer period" in France, is proved by its remains in the caves where men dwelt, and by a drawing cut by a hunter of the older Stone Age on one of its own tusks ! A cast of this most interesting relic may be seen in the prehistoric collection at the British Museum, and shows that the men of that time were not devoid of artistic power (see Fig. 53). Some of the lines in this illustration represent cracks in the original, so that the actual outline is not easily made out. But here we see the head particularly well drawn, the tusks and downward lines indicating the hairy mane. Reindeer and other

FIG. 53.—Figure of the Mammoth, engraved on Mammoth ivory by cave-men, La Madelaine, France. In the Lartet Collection, Paris.

animals were also engraved on horn, etc., by the men who were contemporary with the Mammoth.

We know that man has exterminated a great many noble animals in his time, and, alas ! continues to do so at the present time in Africa, and in North and South America. The giraffe and the bison, once so plentiful, are now almost extinct. Primitive man was a hunter, and, as he multiplied, his wants became greater, and more animals were therefore destroyed. Probably the same explanation applies to the great Moa bird of New Zealand, and possibly even to the Megatherium of South America.

With regard to the tusks of the Mammoth, which are considerably larger than those of either the African or Indian elephant, it is evident that they must have been of some service, for Nature would never have endowed the animal with such great and ponderous instruments—to support which the skull is greatly modified in both the Mammoth and elephant—without some definite purpose. We have often been asked how the Mammoth used his tusks; now, this question can best be answered by reference to the habits of living elephants. The elephant of to-day is a fairly peaceable creature, but, if attacked, can despatch the aggressor in various ways. Some enemies he can crush under his feet; a man he can pick up with his trunk and hurl to a considerable distance, probably with fatal results. But the tusks do not appear to be used as weapons of offence or defence. We must consider how the animal feeds. The general food of the elephant consists of the foliage of trees. In Africa it feeds largely on mimosas. Now, it is clear that, in spite of having a long trunk, an elephant cannot obtain all the leaves of a tall tree while the tree remains standing; mimosa trees, for instance, are often thirty feet high, and have richer foliage at the crown. So it appears that they actually overturn them. On this point we have the testimony of Sir Samuel Baker, who says, "The destruction caused by a herd of elephants in a mimosa forest is extraordinary, and I have seen trees uprooted of so large a size that I am convinced no single elephant could have overturned them. I have measured trees four feet six inches in circumference, and about thirty feet high, uprooted by elephants. The natives assured me that the elephants mutually assist each other, and that several engage together in the work of overturning a large tree. None of the mimosas have tap-roots; thus the powerful tusks of the elephants applied as crowbars at the roots, while others pull at the branches with their trunks, will effect the destruction of a tree so large as to appear invulnerable." Another writer says the elephant also feeds on a variety of bulbs, the situation of which is indicated by his exquisite sense of smell, and that, to obtain these, he turns up the ground with his tusks, so that whole acres may be seen thus ploughed up.

Now, in Siberia, where the ground would be harder, we can imagine that the larger tusks of the Mammoth would be highly serviceable in uprooting fir trees and breaking off their branches, for Benkendorf's fortunate discovery informs us that such trees formed at least part of their food.

CHAPTER XIV.

THE MASTODON AND THE WOOLLY RHINOCEROS.

> "Of one departed world
> I see the mighty show."

ANOTHER elephantine monster, evidently allied to the Mammoth, was the Mastodon, a creature which there is reason to think was contemporary, in America, with the men of a prehistoric age. It was so named by Baron Cuvier to distinguish it from the Mammoth, with which it was by others considered identical; and his discrimination of the two forms marked an important and early step in the history of palæontology. The chief difference between these two extinct types lies in their molar teeth. These, on cutting the gum, must have exhibited a number of somewhat conical protuberances of a mammiform appearance; hence the name.[1] As these points were worn down by mastication, the surface of the tooth showed a series of discs of various sizes. The teeth were covered by a very thick coat of dense, brittle enamel. There are, however, differences in the bony framework of the animal, as well as in its general proportions, which serve to distinguish it from the Mammoth; but it will not be necessary to enter into these matters here, for this is difficult ground, even to the student who is well versed in anatomy. Notwithstanding a vast amount of observation on the subject, considerable differences of opinion have prevailed among palæontologists with

[1] Greek—*mastos*, teat; *odous, odontos*, tooth.

regard to the proper relation of the Mastodon to the Mammoth and living elephants.

At the entrance of the Geological Gallery in the Natural History Museum, South Kensington, the reader will see a magnificent skeleton of an American Mastodon, of which more presently. On this specimen our artist has based his restoration, Plate XXI. A large part of the great gallery referred to is devoted to the fossil remains of proboscideans; that is, creatures provided with a long proboscis, or trunk, such as elephants and Mastodons. This collection, from widely different quarters, is the largest and

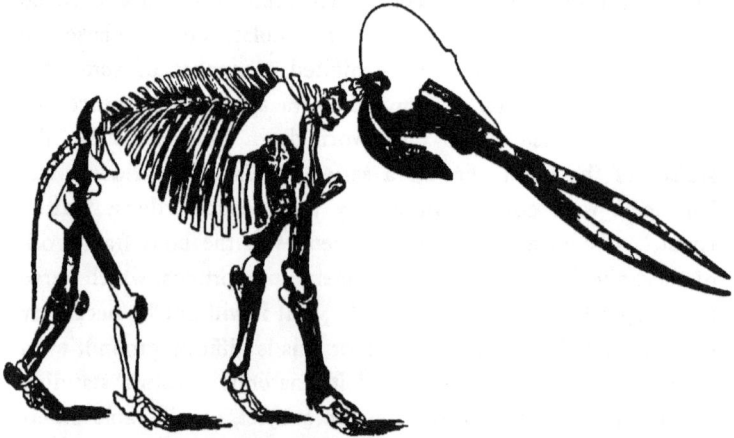

FIG. 54.—Skeleton of *Mastodon arvernensis*, Pliocene, Europe.

most complete in the world. By comparing the specimens of teeth in the cases, and looking at the fine specimens of skulls, and the numerous bones and tusks in the side cases, the reader will carry away a better idea than we can convey by description. Fig. 54 shows the skeleton of Mastodon arvernensis with two very long tusks. Mastodon augustidens had four tusks, two in each jaw, but one of those in the lower jaw sometimes dropped out as the animal grew older.

No genus of quadrupeds has been more extensively diffused over the globe than the Mastodon. From the tropics it has

From Nanaimo in British Columbia there comes the following on the subject of alleged live mastodons :—" The Stickeen Indians positively assert that within the last five years they have frequently seen animals which from descriptions given must be mastodons. Last spring, while out hunting, one of these Indians came across a series of large tracks, each of the size of the bottom of a salt barrel, sunk deep in the moss. He followed the curious trail for some miles, finally coming out in full view of his game. As a class these Indians are the bravest of hunters, but the proportions of this new species of game filled the hunter with terror, and he took to swift and immediate flight. He described the creature as being as large as a post-trader's store, with great, shining, yellowish white tusks, and a mouth large enough to swallow a man at a single gulp. He further says that the animal was undoubtedly of the same species as those whose bones and tusks lie all over this section of the country. The fact that other hunters have told of seeing these monsters browsing on the herbs up along the river gives a certain probability to the story. Over on Forty-mile creek bones of mastodons are quite plentiful. One ivory tusk, nine feet long, projects from one of the sand dunes on the creek, and single teeth have been found that were so large that they would be a good load for one man to carry."

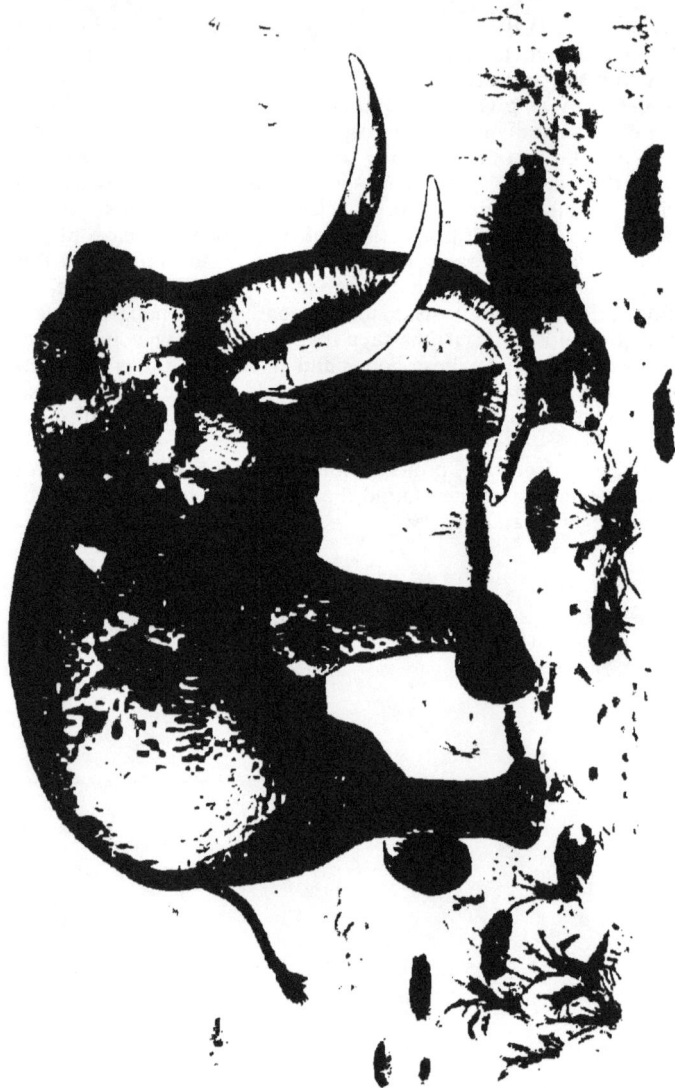

PLATE XXI.   THE MASTODON OF OHIO, M. AMERICANUS.

extended both north and south into temperate regions, and in America its remains have been discovered as high as latitude 66° N. But the true home of the Mastodon giganteus, in the United States, like that of M. augustidens in Europe, lies in a more temperate zone, and, as Professor Owen says, we have no evidence that any species was specially adapted, like the Mammoth, for braving the rigours of an arctic winter.

Now, we know from trustworthy geological evidence that the Mastodon is a much older form of life than the Mammoth. The record of the rocks tells us that it first put in an appearance in an early Tertiary period known as the Miocene (see Table of Strata, Appendix I.), and in the Old World lived on to the end of the succeeding Pliocene period. But in America several species, especially M. giganteus, survived till late in the Pleistocene period, where it was probably seen by primitive men. This is all that is known about its geographical range, and its antiquity or range in time ; some day, perhaps before very long, palæontologists may be able to trace the great proboscideans further back in time, and to show from what form of animal they were derived. Strange as it may seem, anatomists declare that they show some remote affinity with the rodents, or gnawing animals, and, in some respects, even with Sirenians, such as the Manatee (see Chapter XVI.). But at present the evolution of this remarkable group of animals is an unsolved problem. Those strange animals, the Dinocerata, from Wyoming, described in chap. x., may perhaps give some indication as to the direction in which we must look for the elephant's ancestors. We noticed that their limbs were decidedly elephantine (see p. 150), but they had no trunks, and their skulls showed curious prominences like horn-cores ; their teeth too are very different.

The visitor to the Geological Collection at South Kensington will also notice a splendid cranium of an elephant, with very long tusks, from the famous Sivalik Hills of Northern India[1] (Stand

[1] There is some difficulty in determining the precise geological age of the strata in question, on account of the curious mixture of fossil forms of life they contain ; but many authorities consider them to be of older Pliocene age.

D on plan). It belonged to Elephas ganesa, one of the largest of all the fossil elephants known. The total length of the cranium and tusks is fourteen feet, and the tusks alone measure ten feet six inches in length! This remarkable specimen was presented by Sir William Erskine Baker, K.C.B.

But to return to our Mastodon. It was early in the eighteenth century that the teeth and bones of the Mastodon were first described,[1] and it is curious to observe how differently scientific discoveries were regarded in those days; for this society of learned men published in these *Transactions* a letter from Dr. Mather to Dr. Woodward, in which the former gives an account of a large work in manuscript, but does not name the author. This book, which appears to have been a commentary on the Bible, Dr. Mather recommends "to the patronage of some generous Mœcenas to promote the publication of it," and transcribes, as a specimen, a passage announcing the discovery at Albany, now the capital of New York State, in the year 1705, of enormous bones and teeth. These relics he considered to belong to a former race of giants, and appeals to them in confirmation of Genesis vi. 4 ("The Nephilim (giants) were in the earth in those days ").

Portions of the skeleton of Mastodon, discovered in 1801, were sent to England and France, and two complete specimens were at length put together in America. One of these was exhibited as a Mammoth, in Bristol and London, by Mr. C. W. Peale, a naturalist, by whom they were found in marly clay on the banks of the Hudson, near Newburgh, in the State of New York.

Previous to this, in 1739, a French officer, M. de Longueil, traversed the virgin forests bordering on the river Ohio, in order to reach the Mississippi, and the Indians who escorted him accidentally discovered, on the borders of a marsh, various bones, some of which seemed to be those of unknown animals. In this turfy marsh, known as the Big Bone Lick, or Salt Lick, in

---

[1] *Philosophical Transactions of the Royal Society*, 1714, vol. xxix.

consequence of the saltness of its waters, herds of wild animals collect together, attracted by the salt, for which they have a great liking. This is probably the reason why so many bones have accumulated here. M. de Longueil carried away some bones and teeth, and, on his return to France, presented them to Daubenton and Buffon. The former declared the teeth to be those of a hippopotamus, and the tusk and gigantic thigh-bone he reported to belong to an elephant. Buffon, however, did not share this opinion, and succeeded in converting Daubenton, as well as other French naturalists, to his views. He gave to this fossil animal the name of " the Elephant of Ohio," but formed an exaggerated idea of its size.

This discovery produced a great impression in Europe. The English, becoming masters of Canada by the peace of 1763, sought eagerly for more remains. Croghan, the geographer, visited the Big Bone Lick, and found there some more bones of the same kind. He forwarded many cases to different naturalists in London.

Sir Henry Howorth, in his recent work, *The Mammoth and the Flood* (in which are brought forward certain views not shared by most geologists), mentions that in 1762 the Shawnee Indians found, some three miles from the river Ohio, the skeletons of five Mastodons, and reported that one of the heads had a long nose attached to it, below which was the mouth. Several explorers report discoveries of a like nature, which, if they may be trusted, and if they really refer to the Mastodon, and not the Mammoth, seem to show that portions of the skin and hairy covering have been seen. If so, their preservation is probably due to the saltness of the waters of this marsh, for salt is a good preservative. In *The American Journal of Science*,[1] Dr. Koch reports the discovery of a Mastodon's skeleton, of which the head and fore foot were well preserved, also large pieces of the skin, which looked like freshly tanned leather. But some of

[1] Vol. xxxvi. p. 199.

these accounts refer to tufts of hair—in one case three inches long.

The great skeleton of Mastodon americanus already referred to was purchased by the trustees of the British Museum, of Mr. Albert Koch, a well-known collector of fossil remains, who had exhibited it in the Egyptian Hall, Piccadilly, in 1842 and 1843, under the name of " the Missouri Leviathan," an enormous and ill-constructed monster, made up of the bones of this skeleton, together with many belonging to other individuals, in such a way as to horrify an anatomist and appeal all the more forcibly to the imagination of the public. From this heterogeneous assemblage of bones those belonging to the same animal have now been selected and articulated in their proper places. The height of this specimen is nine feet and a half, and the total length about eighteen feet.

According to Mr. Koch, the remains exhibited by him were found in alluvial deposits on the banks of a small tributary of the Osage River, in Benton County, Missouri. The bones were embedded in a brown, sandy deposit, full of· vegetable matter, in which were recognised remains of the cypress, tropical cane, swamp moss, etc., and this was covered by blue clay and gravel to a thickness of about fifteen feet. Mr. Koch personally assured Dr. Mantell that an Indian flint arrow-head was found beneath the leg-bones of this skeleton, and that four similar weapons were embedded in the same stratum. He declared that he took them out of the bed with his own hands.

In the Pier-case (No. 38), near the Mastodon americanus, may be seen fifteen heads and jaws, together with other parts of the skeleton, mostly obtained from the same locality, but some of them came from the " Big Bone Lick," Kentucky.

A fine specimen, obtained from a marsh near Newburgh, by Dr. Warren, measured eleven feet in height, and seventeen in length, while the tusks were nearly ten feet long, not including the portion in the long sockets of the cranium. Twenty-six

An interesting find was that of Dr. Barton, a professor of the University of Pennsylvania. At a depth of six feet, and under a great bank of chalk, bones of the Mastodon were found sufficient to form a skeleton, and in the middle of the bones was seen a mass of vegetable matter enveloped in a kind of sac (which probably was the stomach of the animal). This matter was found to be composed of small leaves and branches, amongst which was recognised a species of rush yet common in Virginia. In North America, where the Mastodon survived into the period of primitive man, various strange legends exist that seem to refer to it. Traditions were rife among the Red Men concerning this giant form and its destruction.

A French officer named Fabri informed M. Buffon, the naturalist, that the "savages" (Indians) regarded the bones found in various parts of Canada and Louisiana as belonging to an animal which they named "Father of the Ox." The Shawnee Indians believed that with this enormous animal there existed men of proportionate development, and that the Great Being destroyed both with thunderbolts. Those of Virginia state that as a troop of these terrible animals were destroying the deer, bisons, and other animals created for the use of Indians, the Great Man slew them all with his thunder, except the Big Bull, who shook off the thunderbolts as they fell on him, till at last, being wounded in the side, he fled towards the great lakes, where he lies to this day.

This is one of the songs which Fabri heard in Canada: "When the great *Manitou* descended to the earth, in order to satisfy himself that the creatures he had created were happy, and he interrogated all the animals, the bison replied that he would be quite contented with his fate in the grassy meadows, where the grass reached his belly, if he were not also compelled to keep his eyes constantly turned towards the mountains to catch the first sight of the 'Father of the Ox,' as he descended, with fury, to devour him and his companions." Many other tribes repeat similar legends.

The bones with which Mazuyer practised his famous deception were those of a Mastodon (see p. 196).

Contemporary with the Mammoth in Siberia and in Northern and Western Europe, was the "Woolly Rhinoceros" (Rhinoceros tichorhinus). Its body has been found in frozen soil in Siberia, with the skin, the two horns, the hair, and even the flesh preserved, as in the case of the Mammoth. It had a smooth skin without folds, covered with a fine curly and coarse hairy coat, to enable it to withstand the rigours of an arctic climate. The

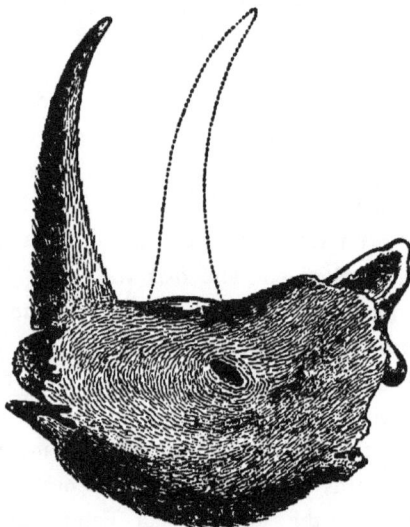

FIG. 55.—Head of Woolly Rhinoceros, partly restored by M. Deslongchamps.

traveller Pallas gives a long account of one of these creatures, which was taken out of the ice, with its skin, hair, and flesh preserved. The following is a brief summary of his narrative. The body was observed in December, 1771, by some Jakuts near the river Vilui, which discharges itself into the Lena below Jakutsk in Siberia, latitude 64° north. It lay in frozen sand upon the banks of the river. A certain Russian inspector had sent on to Irkutsk

THE WOOLLY RHINOCEROS, RHINOCEROS TICHORHINUS.
Cotemporary with the Mammoth.

PLATE XXII.

the head and two feet of the animal, all well preserved. The rest of it was too much decomposed, and so was left. The head was quite recognisable, since it was covered with its leathery skin. The eyelids had escaped total decay (see Fig. 55). The skin and tendons of the head and feet still preserved considerable flexibility. He was, however, compelled to cross the Baikal lake before the ice broke up, and so could neither draw up a sufficiently careful description nor make sketches of those parts which were sufficiently preserved. Plate XXII. is a restoration.

The rhinoceros in question was neither large for its species nor advanced in age; but it was at least fully grown. The horns were gone, but had left evident traces on the head. The skin which covered the orbits of the eyes and formed the eyelids was so well preserved, that the openings of the eyelids could be seen, though deformed and scarcely penetrable to the finger. The foot that was left—after some parts had unfortunately been burned while left to dry slowly on the top of a furnace—was furnished with hairs. These hairs adhering in many places to the skin, were from one to three lines in length, tolerably stiff and ash-coloured. What remained proved that the foot was covered with bunches of hair hanging down.

Like the Mammoth and the Mastodon, its contemporaries, the Woolly Rhinoceros has given rise to some curious legends. In the city of Klagenfurt, in Carinthia, is a fountain on which is sculptured the head of a monstrous dragon with six feet, and a head surmounted by a stout horn. According to popular tradition, still prevalent at Klagenfurt, this dragon lived in a cave, whence it issued from time to time to frighten and ravage the country. A bold cavalier killed the dragon, paying with his life for this proof of courage. The same kind of legend seems to be current in every country, such as that of the valiant St. George and the dragon, and of St. Martha, who about the same time conquered the famous *Tarasque* of the city of Languedoc, which bears the name of Tarascon.

Q

But at Klagenfurt the popular legend has happily found a mouthpiece; the head of the pretended dragon killed by the valorous knight is preserved in the Hôtel de Ville, and this head has furnished the sculptor of the fountain with a model for the head of his statue. Herr Unger, of Vienna, recognised at a glance the cranium of the fossil rhinoceros; its discovery in some cave had probably originated the fable of the knight and the dragon. It is always interesting to discover a scientific basis for fables which otherwise it would be difficult to account for.

The same rhinoceros was once a denizen of our country, and its remains are met with in caves and river-gravels. Specimens of its skull have also been dredged up by fishermen from the " Dogger Bank" in the North Sea.

CHAPTER XV.

"To discover order and intelligence in scenes of apparent wildness and confusion is the pleasing task of the geological inquirer."—DR. PARIS.

OF all the monsters that ever lived on the face of the earth, the giant birds were perhaps the most grotesque. An emu or a cassowary of the present day looks sufficiently strange by the side of ordinary birds; but "running birds" much larger than these flourished not so very long ago in New Zealand and Madagascar, and must at one time have inhabited areas now sunk below the ocean waves.

The history of the discovery of these remarkable and truly gigantic birds in New Zealand, and the famous researches of Professor Owen, by which their structures have been made known, must now engage our attention.

In the year 1839 Professor Owen exhibited, at a meeting of the Zoological Society, part of a thigh-bone, or femur, 6 inches in length, and $5\frac{1}{2}$ inches in its smallest circumference, with both extremities broken off. This bone of an unknown struthious bird was placed in his hands for examination, by Mr. Rule, with the statement that it was found in New Zealand, where the natives have a tradition that it belonged to a bird now extinct, to which they give the name Moa. Similar bones, it was said, were found buried on the banks of the rivers.

A minute description of this bone was given by the professor, who pointed out the peculiar interest of this discovery on account of the remarkable character of the existing fauna of New Zealand, which still includes one of the most extraordinary birds of the struthious order ("running birds"), viz. the Apteryx, and also because of the close analogy which the event indicated by the present relic offers to the extinction of the Dodo in the island of Mauritius. On the strength of this one fragment he ventured to assert that there once lived in New Zealand a bird as large as the ostrich, and of the same order. This conclusion was more than confirmed by subsequent discoveries, which he anticipated; and, as we shall see, his estimate was a most moderate one, for the extinct bird turned out to be considerably larger than the ostrich.

Later on he received from a friend in New Zealand news of the discovery of more bones. In 1843 a collection of bones of large birds was sent to Dr. Buckland, Dean of Westminster, by the Rev. William Williams, a zealous and successful Church missionary, long resident in New Zealand. On sending off his consignment Mr. Williams wrote a letter, of which we give the greater part below.

" Poverty Bay, New Zealand, February 28, 1842.
" DEAR SIR,
    " It is about three years ago, on paying a visit to this coast—south of the East Cape, that the natives told me of some extraordinary monster, which they said was in existence in an inaccessible cavern on the side of a hill near the river Wairoa; and they showed me at the same time some fragments of bone taken out of the beds of rivers, which they said belonged to this creature, to which they gave the name Moa.
    " When I came to reside in this neighbourhood I heard the same story a little enlarged; for it was said that this creature *was still existing* at the said hill, of which the name is Wakapunake, and that it is guarded by a reptile of the lizard species [genus]; but I could not learn that any of the present generation had seen it. I still considered the whole as an idle fable, but offered

a large reward to any one who would catch me, either the bird or its protector. . . ."

These offers procured the collection of a considerable number of fossil bones, on which Mr. Williams, in his letter, makes the following observations :—

"None of these bones have been found on the dry land, but are all of them from the banks and beds of fresh-water rivers, buried only a little distance in the mud. . . . All the streams are in immediate connection with hills of some altitude.

"2. This bird was in existence here at no very distant time; though not in the memory of any of the inhabitants; for the bones are found in the beds of the present streams, and do not appear to have been brought into their present situation by the action of any violent rush of waters.

"3. They existed in considerable numbers"—an observation which has since been abundantly confirmed.

"4. It may be inferred that this bird was long-lived, and that it was many years before it attained its full size." This is doubtful.

"5. The greatest height of the bird was probably not less than fourteen or sixteen feet." Fourteen is probably the extreme limit.

"Within the last few days I have obtained a piece of information worthy of notice. Happening to speak to an American about these bones, he told me that the bird is still in existence in the neighbourhood of Cloudy Bay, in Cook's Straits. He said that the natives there had mentioned to an Englishman belonging to a whaling party that there was a bird of extraordinary size to be seen only at night, on the side of a hill near the place, and that he, with a native and a second Englishman, went to the spot; that, after waiting some time, they saw the creature at a little distance, which they describe as being about fourteen or sixteen feet high. One of the men proposed to go nearer and shoot, but his companion was so exceedingly terrified, or perhaps both of them, that they were satisfied with looking at the bird, when, after a little time it took alarm, and strode off up the side of the mountain.

"This incident might not have been worth mentioning, had it not been for the extraordinary agreement in point of size of the bird "—with his deductions from the bones. " *Here* are the bones which will satisfy you that such a bird *has been* in existence ; and *there* is said to be the *living bird,* the supposed size of which, given by an independent witness, precisely agrees." In spite, however, of several tales of this kind, it is almost certain that these birds are now quite extinct.

The leg-bones sent to London greatly exceeded in bulk those of the largest horse. The leg-bone of a tall man is about 1 ft. 4 in. in length, and the thigh of O'Brien, the Irish giant, whose skeleton, eight feet high, is mounted in the Museum of the Royal College of Surgeons, is not quite two feet. But some of the leg-bones (tibiæ) of Moa-birds measure as much as 39 inches.

In 1846 and 1847 Mr. Walter Mantell, eldest son of Dr. Mantell, who had resided several years in New Zealand, explored every known locality within his reach in the North Island. He also went into the interior of the country and lived among the natives for the purpose of collecting specimens, and of ascertaining whether any of these gigantic birds were still in existence ; resolving, if there appeared to be the least chance of success, to penetrate into the unfrequented regions, and obtain a live Moa. The information gathered from the natives offered no encouragement to follow up the pursuit, but tended to confirm the idea that this race of colossal bipeds was extinct. He succeeded, however, in obtaining a most interesting collection of the bones of Moa-birds, belonging to birds of various species and genera, differing considerably in size. This collection was purchased by the trustees of the British Museum for £200. Another collection was made by Mr. Percy Earle from a submerged swamp, visible only at low water, situated on the south-eastern shore of the Middle Island. This collection also was purchased by the trustees for the sum of £130. Mr. Walter Mantell, who described this locality, near Waikouaiti, seventeen miles north of Otago, thinks it was originally a swamp or morass, in which the New Zealand flax once grew luxuriantly. The appearance and position of the bones are similar to those of the quadrupeds embedded in peat-bogs, as, for instance, the great Irish elk (see next chapter). They have acquired a rich umber colour, and their texture is firm and tough. They still contain a large proportion of animal matter. Unfortunately, even when Mr. Walter Mantell visited this spot, the bed containing the bones was rapidly diminishing from the inroads of the sea, and perhaps by this time is entirely washed away. Mr. W. Mantell, however, obtained fine specimens and feet of a large Moa-bird (Dinornis) in an upright position ; and there seems to be little doubt that the unfortunate bird was mired in the swamp, and perished on the spot.

The bones which he obtained from the North Island presented a different appearance, being light and porous, and of a delicate fawn-colour. They were embedded in loose volcanic sand. Though perfect, they were as soft and plastic as putty, and required most careful handling. They were dug out with great care, and exposed to the air and sun to dry before they could be packed up and removed.

The natives' were a great source of trouble to him, for as soon as they caught sight of his operations they came down in swarms—men, women, and children, trampling on the bones he had laid out to dry, and seizing on every morsel they could get. The reason of this was that their cupidity and avarice had been excited by the large rewards given by Europeans in search of these treasures. Mixed with the bones he found fragments of shells, and sometimes portions of the windpipe, or trachea.

One portion of an egg which he found was large enough to

enable him to calculate the size of the egg when complete. " As a rough guess, I may say that a common hat would have served as an egg-cup for it : what a loss for the breakfast-table ! And if many native traditions are worthy of credit, the ladies have cause to mourn the extinction of the Moa : the long feathers of its crest were by their remote ancestors prized above all other ornaments ; those of the White Crane, which now bear the highest value, were mere pigeon's feathers in comparison."

The total number of species of Moa once inhabiting New Zealand was probably at least fifteen, and, judging from the enormous accumulations of their bones found in some districts, they must have been extremely common, and probably went about in flocks. " Birds of a feather *flock* together " (proverb).

It is justly concluded, both from the vast number of bones discovered, and from the fact of their great diversity in size and other features, that they must have had the country pretty much to themselves ; or, in other words, they enjoyed immunity from the attacks of carnivorous quadrupeds. In whatever way the Moas originated in New Zealand, it is evident that the land was a favourable one, for they multiplied enormously, and spread from one end to the other. Not only was the number of individuals very large, but they belonged (according to Mr. F. W. Hutton) to no less than seven genera, containing twenty-five different species, a remarkable fact which is unparalleled in any other part of the world. The species described by Professor Owen in his great work,[1] vary in size from 3 ft. to 12 or even 14 ft. in height, and differ greatly in their forms, some being tall and slender, and probably swift-footed like the ostrich, whilst others were short and had stout limbs, such as Dinornis elephanto-

[1] Memoir on *The Extinct Wingless Birds of New Zealand.* London, 1878. The beautiful drawing by Mr. Smit (Plate XXIV.) is from a photograph in this valuable work representing the late Sir Richard Owen standing in his academic robes by the side of a specimen of the skeleton of the great Dinornis maximus.

MOA-BIRDS.

Dinornis giganteus.   *D. elephantopus.*

PLATE XXIII.

pus (Fig. 56), which was undoubtedly a bird of great strength, but very heavy-footed. Dinornis crassus also had stout limbs. (See Plate XXIII.)

The Natural History Museum at South Kensington contains a valuable collection of remains of Moa-birds. These skeletons may

FIG. 56.—*A*. Skeleton of the Elephant-footed Moa, *Dinornis elephantopus*, from New Zealand. *B*. Leg-bones of *Dinornis giganteus*, representing a bird over 12 ft. high. *r*, *b*, footprints.

be seen in Gallery No. 2 (at the end of the long gallery) in the glass cases R, R', and S. Dinornis elephantopus (elephant-footed) is in front of the window. In D. giganteus the leg-bone (see Fig. 56) attains the enormous length of 3 ft., and in an allied species it is even 39 in. ! The next bone below (cannon bone) is sometimes more than half the length of the leg-bone (tibia).

A skeleton in one of the glass cases has a height of about 10½ ft., and it is concluded that the largest birds did not stand less than 12 ft., and possibly were 14 ft. high !

Dinornis parvus (the dwarf Moa) was only three feet high.

In 1882 the trustees obtained, from a cave in Otago, the head, neck, two legs, and feet of a Moa (D. didinus), having the skin, still preserved in a dried state, covering the bones, and some few feathers of a reddish hue still attached to the leg (Table case 12). The rings of the windpipe may be seen *in situ*, the sclerotic plates of the eye, and the sheaths of the claws. One foot also shows the hind claw still attached.

From traditions and other circumstances it is supposed that the present natives of New Zealand came there not more than about six hundred years ago, and there is reason to believe that the ancient Maoris, when they landed, feasted on Moa-birds as long as any remained. Their extermination *probably* only dates back to about the period at which the islands were thrice visited by Captain Cook, 1769–1778. The Moa-bird is mixed up with their songs and stories, and they even have a tradition of caravans being attacked by them. Still, some people believe that they were killed off by the race which inhabited New Zealand before the Maoris came. But they must have been there up to a time not far removed from the present. It is even said that the "runs" made by them were visible on the sides of the hills up to a few years ago ; and possibly they may still be visible. The charred bones and egg-shells have been found mixed with charcoal where the native ovens were formerly made, and their eggs are said to have been found in Maori graves. Mr. Hutton considers that in the North Island they were exterminated three or four centuries ago, while in the South Island they may have lingered a century longer.

The nearest ally of the Moa is the small Apteryx, or Kiwi, of

New Zealand, specimens of which may be seen at the Natural History Museum, at the end of the long gallery devoted to living birds. This bird, however, has a long pointed bill for probing in the soft mud for worms, whereas the bill of the Moa was short like that of an ostrich.

Another difference between the two is that, while the Kiwi still retains the rudiments of wing-bones, the Moa had hardly a vestige of such.

In Australia the remains have been found of a bird probably related to the Cassowaries, but at present imperfectly known. To this type of struthious, or running bird, the name Dromornis has been given.

Now, it is a remarkable fact that remains of another giant bird and its eggs have been found on the opposite side of the great Indian Ocean, namely, in the island of Madagascar, the existence of which was first revealed by its eggs, found sunk in the swamps, but of which some imperfect bones were afterwards discovered. One of these eggs was so enormous that its diameter was nearly fourteen inches, and was reckoned to be as big as three ostrich eggs, or 148 hen's eggs! This means a cubic content of more than two gallons! The natives search for the eggs by probing in the soft mud of the swamps with long iron rods. A large and perfect specimen of an egg of this bird, such as was recently exhibited at a meeting of the Zoological Society, is said to be worth £50. What the dimensions of Æpyornis were it is impossible to say, and it would be unsafe to venture a calculation from the size of the egg.[1] The reader who wishes to see some of the remains of this huge bird may be referred to the Natural History Museum. In wall case No. 25, Gallery 2 (Geological Department), may be seen a tibia and plaster casts of other bones; also two entire eggs, many broken pieces, and one

---

[1] From the size of a femur and tibia of *Æpyornis* preserved in the Paris Museum, it could not have been less in stature than the Dinornis elephantopus of New Zealand.

plaster cast of an egg found in certain surface deposits in Madagascar. In the same case may be seen bones of the Dodo from the isle of Mauritius. Unlike New Zealand, Madagascar possesses no living wingless bird. But in the neighbouring island of Mauritius the Dodo has been exterminated less than three centuries ago. The little island of Rodriguez, in the same geographical province, has also lost its wingless Solitaire.

It will thus be seen that we have three distinct groups of giant land birds—the Moas, the Dromornis, and the Æpyornis,—occupying areas at present widely separated by the ocean.

This raises the difficult but very interesting question, how they got there; and the same applies to their living ancestors. The ostrich proper, Struthio camelus, inhabits Africa and Arabia; but there is evidence from history to show that it formerly existed in Beluchistan and Central Asia. And, going still further back, the geological record informs us that, in the Pliocene period, they inhabited what is now Northern India. In Australia we have the Cassowary (Casuarius) and the Emeu (Dromaius); in New Zealand, the Apteryx (or Kiwi). Now, as none of these birds can either fly or swim, it is impossible that they could have reached these regions separated as they now are; and it is hardly likely that they arose spontaneously in each district from totally different ancestors. But the new doctrine of evolution affords a key to the problem, and tells us that they all sprang from a common ancestor, of the struthious type (probably inhabiting the great northern continental area), and gradually migrated south along land areas now submerged. In this way we get some idea of the vast changes that have taken place in the geography of the world during later geological periods. Perhaps they were compelled to move south until they reached abodes free from carnivorous enemies. Having done so, they evidently flourished abundantly, especially in New Zealand, where there are so few mammals, except those recently introduced by man.

In North America Professor Cope has reported a large wingless fossil bird from the Eocene strata of New Mexico. In England we have two such—namely, the Dasornis, from the London Clay of Sheppey (Eocene period), and the Gastornis, from the Woolwich beds near Croydon, and from Paris (also Eocene).

It will thus be seen that big struthious birds have a long history, going far back into the Tertiary era, and that they once had a much wider geographical range than they have now. Doubtless, future discoveries will tend to fill up the gaps between all these various types, both living and extinct, and to connect them together in one chain of evolution.

The last great find of Moa-birds in New Zealand took place only last year, and was reported by a correspondent to the *Scotsman* (November 13, 1891), writing from Oamaru. In the letter that appeared at the above date, our friend Mr. H. O. Forbes announces the discovery of an immense number of bones, estimated to represent at least five hundred Moas! They were found in the neighbourhood of Oamaru. And, after some preliminary remarks, he continues as follows :—

" The part of the field on which the remains were found bears no traces of any physical disturbance—*e.g.* of earthquake, or flood, or hurricane—that would account for the sudden destruction of a flock or 'mob' of Moas. The Moa, when alive, carried in its crop—like our own hens—a quantity of stones to serve as a private coffee-mill for digestive grinding; stones which, being somewhat in proportion to the magnitude of the giant bird, form, when found in one place, a 'heap' of stones which are easily identified as a Moa heap, and nothing else. And in the present case the heap was here and there found in such relation to the bones of an individual bird as to show that the Moa must have died on that spot, and remained there quietly undisturbed. Further, the number of birds represented by the exhumed remains is so great that the living birds could not have stood together on the space of ground on which the remains were found lying. And there is not on any of the bones any trace of such violence as must have left its mark if the death of the birds had been caused by a Moa-hunting mankind. Finally, it does not appear that in this particular district there ever has been, at any traceable period of the physical history of the land, a forest vegetation, such as might suggest that the catastrophe was caused by fire.

" The question how to account for the slaughter is raised like-wise by two previous finds of Moa bones. The first of these, at Glenmark, in Canterbury, was the most memorable, because, being the first, it made the deepest impression. The second great find, far inland, up the Molineux River, otherwise the Clutha, was beneath the diluvium that is now worked by the gold-digger. The spot must have been the site of a lagoon, at one point of which there was a spring. Round about this point there were found the remains of, it was reckoned, five hundred individual Moas. The bones were quietly *laid* there, with, in some cases, the 'heap' of digestive stones *in situ* along with the skeletons. And Mr. Booth, whose elaborate investiga-tion of this case is recorded in the annual volume of *The New Zealand Institute*, suggested the theory that the climate of New Zealand was changing to a degree of cold intolerable to Moa nature ; and that the birds, fleeing from its rigour, sought comfort in the spring of water, sheltering their featherless breast in it, and so dozing out of this troubled life. And in this new find the wonder comes back unmitigated, as a mystery unsolved. For here is no bog deep enough, as in the first instance, nor lagoon spring, as in the second, to account for that multitude of giant birds dying in one spot.

" Another curious puzzle is, on close inspection, found every-where in the Moa bone discoveries. It is hardly possible to make sure that the bones of any one complete Moa skeleton all belong to the same individual I heard some one say the other day that it is not certain that any Moa in any earthly museum has all his own bones, and only his own.

" A main interest of such a find lies not in the power of supply-
ing museums with specimens of what is rapidly disappearing from
the face of the world, but in the possibility of finding species
of Moa that have not hitherto been tabulated. Whether any
new species have been brought to light on this occasion the
experts will not say until there has been time to make a careful
study of the bones, nor do they venture on any theory to account
for there being so many individual birds dead in that one place,
where there appears to be no room for the explanations offered
in connection with previous great finds. The date of these birds
appears to be earlier than that of the coming of the Maoris into
New Zealand, say five or six hundred years ago, as the Maori
memory appears to have in it no trace of feasting on these giant
Moas, but celebrates the rat-hunt in its ancient heroic song. And
your readers may picture their appearance by noticing the fact
that one of the recently found bones must have belonged to a
Moa fourteen feet high ! "

NOTE.—For further information on this interesting subject, the reader is
referred to a paper in *Natural Science*, October, 1892, by Mr. F. W. Hutton.
In a valuable paper, read before the Royal Geographical Society by Mr. H. O.
Forbes, March 13, 1893, the lecturer alluded to the important fact that bone
belonging to big extinct struthious birds have been discovered in Patagonia.
This is interesting news as bearing upon the theory of a former Antarctic continent connecting Australia and New Zealand with South Africa, and perhaps
even with South America. After the lecture, to which we listened with great
interest, the subject was discussed by Mr. Slater, Dr. Günther, and Dr.
Henry Woodward. For ourselves we can see no great difficulty in accepting
the theory that such a continent once existed, though it is out of harmony
with the now rather fashionable theory of "the permanence of ocean basins"
—a doctrine which seems to have been pressed too far.

CHAPTER XVI

" And, above all others, we should protect and hold sacred those types, Nature's masterpieces, which are first singled out for destruction on account of their size, or splendour, or rarity, and that false detestable glory which is accorded to their most successful slayers. In ancient times the spirit of life shone brightest in these; and when others that shared the earth with them were taken by death they were left, being more worthy of perpetuation. Like immortal flowers they have drifted down to us on the ocean of time, and their strangeness and beauty bring to our imaginations a dream and a picture of that unknown world, immeasurably far removed, where man was not : and when they perish, something of the gladness goes out of nature, and the sunshine loses something of its brightness."—W. H. HUDSON, in *The Naturalist in La Plata*.

AMONG the extinct animals of prehistoric times the "Great Irish Elk,"[1] as it is generally called, deserves special notice, both from the enormous size of its antlers, and from the fact that its remains are exceedingly plentiful in Ireland.

This magnificent creature, so well depicted by our artist (Plate XXV.), was, however, by no means confined to Ireland ; its remains are found in many parts of Great Britain, particularly in cave deposits, and also on the Continent. Some writers think that it was contemporary with men in Ireland ; it may have been so, but at present the question cannot be considered as proved. Mr. R. J. Ussher, who found its remains in a cave near Cappagh, Cappoquin, thinks he has obtained evidence to show that it was

---

[1] The term "Elk" is misleading, for it is not an elk (*alces*) at all, but a true *Cervus* (stag). It should be called "the Great Irish Deer."

hunted by man at the time when he hunted reindeer in this part of Europe, but the age of the strata containing the remains is doubtful. Again, there is a rib in the Dublin Museum with a perforation which is sometimes taken to be the result of a wound from a dart, arrow, or spear; but the wound may have been inflicted by one of the sharp tynes in a fight between two bucks.

Dr. Hart mentions the discovery of a human body in gravel, under eleven feet of peat, soaked in bog-water, in good preservation, and completely clothed in antique garments of hair, which it has been conjectured might be that of the Great Deer. But if some individual animal had perished and left its body under the like circumstances, its hide and hair ought equally to have been preserved. Dr. Molyneux, to whom we owe the first account of its discovery, says that its extinction in Ireland has occurred " so many ages past, as there remains among us not the least record in writing, or any manner of tradition, that makes so much as mention of its name ; as that most laborious inquirer into the pretended ancient but certainly fabulous history of this country, Mr. Roger O'Flaherty, the author of *Ogygia*, has lately informed me." [1]

In the romance of the " Niebelungen," now immortalised by Wagner, which was written in the thirteenth century, the word *shelch* occurs, and is applied to one of the beasts slain in a great hunt a few hundred years before that time in Germany. This word has been cited by some naturalists as probably signifying the Great Irish Deer. But this is mere conjecture, and the word might apply to some big Red Deer. The total silence of Cæsar and Tacitus respecting such remarkable animals renders it highly improbable that they were known to the ancient Britons.

Two entire skeletons of the male, with antlers measuring a little over nine feet from tip to tip, and one skeleton of the hornless doe, are to be seen set up in the middle of the long gallery No. 1 at the Natural History Museum. The drawing in Fig. 57 is from

---

[1] *Philosophical Transactions*, vol. xix. p. 490.

R

a specimen in the Museum of the Royal College of Surgeons (Lincoln's Inn Fields). The height of this specimen to the summit of the antlers is 10 ft. 4 in. The span of the antlers, from tip to tip, is 8 ft. (in the living Moose it is only 4 ft). The

FIG. 57.—Skeleton of Great Irish Deer, *Cervus giganteus*, from shell-marl beneath the peat, Ireland. Antlers over 9 feet across.

weight of the skull and antlers together is 76 lbs., but those of another specimen belonging to the Royal Dublin Society weigh 87 lbs. This great extinct deer surpassed the largest Wapite (Cervus Canadensis) in size, and its antlers were very much larger,

wider, and heavier. In some cases the antlers have measured more than 11 ft. from tip to tip. The body of the animal, as well as its antlers, were larger and stronger than in any existing deer. The limbs are stouter, as might be expected from the great weight of the head and neck. Another and more striking feature is the great size of the vertebræ of the neck; this was necessary in order to form a column capable of supporting the head and its massive antlers. (See Plate XXV.)

The first tolerably perfect skeleton was found in the Isle of Man, and presented by the Duke of Athol to the Edinburgh Museum. It was figured in Cuvier's *Ossemens Fossiles.* Besides those already mentioned at South Kensington and Dublin, there is one in the Woodwardian Museum at Cambridge.

It cannot be doubted that, like all existing deer, the animal shed its antlers periodically, and such shed antlers have been found. When it is recollected that all the osseous matter of which they are composed must have been drawn from the blood carried along certain arteries to the head, in the course of a few months, our wonder may well be excited at the vigorous circulation that took place in these parts.

In the Red Deer the antlers, weighing about 24 lbs., are developed in the course of about ten weeks; but what is that compared to the growth of over 80 lbs. weight in some three or four months?

It is a mistake to suppose that the remains discovered in Ireland were found in peat; they occur not in the peat, but in shell-marls and in clays *under the peat.* This is an important point. For if the remains *were* found in the peat, they would prove that the Great Deer survived into a later period; instead of being (as is believed from geological evidence) contemporary with the Mammoth and Woolly Rhinoceros in this country, and then disappearing from view. As already stated, it existed on the Continent, and may there have been exterminated by man.

Mr. W. Williams, who has explored several peat-bogs in Ireland, marking the site of ancient lakes, and found many specimens in beds underlying the peat, has given much interesting information bearing upon the question of the period when the Great Deer inhabited Ireland, and the manner in which it was preserved in the lake-beds.[1] He spent ten weeks in 1876–77, excavating deer remains in the bog of Ballybetagh, and subsequently made similar excavations in the counties of Mayo, Limerick, and Meath. These peat-bogs occupy the basins of lakes, the deeper hollows of which have long since been silted up with marls, clays, and sands, and in this silt, or mud, the plants which produced the peat grew. In all the bogs examined he found a general resemblance in the order of succession of the beds, with only slight variations in the nature of the materials such as might be easily accounted for by differences in the surrounding rocks. In these deposits the geologist may read, as in a book, the successive changes in climate that have taken place since the time when the country was deeply covered with snow and ice during the Glacial period.

He found at the bottom of the old Ballybetagh Lake, and resting on the true Boulder Clay (a product of the ice-sheet), a fine stiff clay which seems to have been brought in by the action of rain washing fine clay out of the Boulder Clay, that nearly covered the land, and depositing it in the lake. This action probably took place during a period of thaw, when the climate was damp, from the melting of so much ice, and the rainfall considerable. Then the climate improved, the cold of the Glacial period passed away, and the climate became warm. During this phase the next stratum was formed, consisting chiefly of vegetable remains. The summers must have been unusually warm, dry, and favourable to the growth of vegetation on the bed of the lake. About this time the Great Irish Deer appeared on the scene, for its remains were found resting on this layer, or stratum,

[1] *Geological Magazine,* new series, vol. viii. (1881), p. 354.

in a brownish clay. This deposit also was the product of a time when the climate was mild. It is a true lake-sediment, with seams of clay and fine sand, the latter having been brought down by heavy rainfall from the hills, just as at the present day.

Now, we have to consider how these Great Deer got buried in this deposit. How did they get drowned? They may have gone into the lakes to escape from wolves, or possibly to escape from ancient Britons (but that is still doubtful). Perhaps they went into the water to wallow, as is usual with deer, or they may have ventured to swim the lakes (see p. 19). To enter the lake from a sandy shore would be easy enough, but, on reaching the other side, they might find a soft mud instead, into which their small feet would sink; and the more they plunged and struggled, the worse became their plight, until at last, weary and exhausted, the heavy antlers pressed their heads down under the water, and they were drowned.

It does not follow, according to this theory, that either the entire animal ought to be found, or even its legs, sticking in the clay. For a few days it might remain so, but the motion of the waters of the lake would sway the body to and fro, while the gases due to decomposition would render it buoyant, and perhaps raise it bodily off the bottom. Then it might float before the wind, its head hanging down, till it reached the lee-side of the lake. Then the antlers would get fastened in mud near the shore, thus mooring the body until at last so much of the flesh of the neck had decayed that the body got separated from it, leaving the head and antlers near the shore.

Nearly a hundred heads had been found in this lake previous to Mr. Williams's explorations, and yet scarcely six skeletons. At first it is somewhat puzzling to account for this scarcity of skeletons compared with heads; but very likely the bodies, minus their heads, were carried right out of the lake, down a river, and perhaps reached the sea or got stranded somewhere down the river in such a way that the bones were never covered up. But in the Limerick bogs heads and skeletons were often found together. In that district the lakes were probably shallow and with but a feeble current, and so the body never floated away. This explanation by Mr. Williams seems satisfactory.

He reports that the female skulls were rarely met with. Either they were more timid in swimming lakes, or, having no antlers, they may have succeeded in getting out, or the care of their young ones may have kept them out of the lakes during the summer months. The clay in which the remains occur is succeeded by another bed of pure clay, which *never* yields any skulls or bones. This, Mr. Williams thinks, was deposited at a time when the climate was more or less severe, and the musk-ox, reindeer, and other arctic animals came down from more northern regions, even down to the south of France. He concludes that this period marks the extinction of the Great Deer in Ireland, whether rightly or wrongly it is hard to say. Many observers are inclined to think that it lived on to a later period. An interesting fact, having some bearing on the question, is this: that the bones in some cases even yet retain their marrow in the state of a fatty substance, which will burn with a clear lambent flame. Evidence such as this seems to point to a more recent date for its extinction.

STELLER'S SEA-COW.[1]

The Sirenia of the present day form a remarkable group of aquatic herbivorous animals, really quite distinct from the Cetacea (whales and dolphins), although sometimes erroneously classed with them. In the former group are the Dugong and the Manatee. These creatures pass their whole life in the water, inhabiting the shallow bogs, estuaries, and lagoons, and large rivers, but never venturing far away from the shore. They browse

[1] For fuller information, see the *Geological Magazine*, decade iii. vol. ii. p. 412. Paper by Dr. Henry Woodward, F.R.S,

beneath the surface on aquatic plants, as the terrestrial herbivorous mammals feed upon the green pastures on land.

Not a few of the tales of mermen and mermaids owe their origin to these creatures, as well as to seals, and even walruses. The Portuguese and Spaniards give the Manatee a name signifying "Woman-fish," and the Dutch call the Dugong the "Little Bearded Man." A very little imagination, and a memory only for the marvellous, doubtless sufficed to complete the metamorphosis of the half-woman, or man, half-fish, into a siren, a mermaid, or a merman. Hence the general name Sirenia.

The Manatee (*Manatus*) inhabits the west coast and rivers of tropical Africa, and the east coast and rivers of tropical America, the West Indies, and Florida.

The Dugong (*Halicore*) extends along the Red Sea coasts, the shores of India, and the adjacent islands, and goes as far as the northern and eastern coasts of Australia.

The most remarkable Sirenian is the Rhytina gigas, or "Steller's Sea-Cow." Early in 1885 the trustees of the British Museum acquired a nearly complete skeleton of this animal, now extinct, from peat deposits in Behring's Island, of Pleistocene

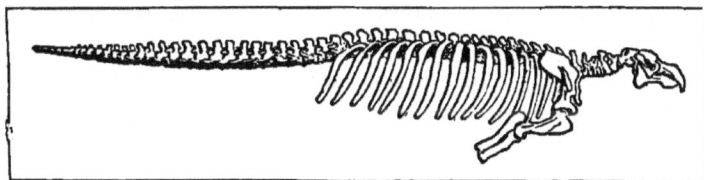

FIG. 58.—Skeleton of *Rhytina gigas* (Steller's "Sea-Cow"), from a peat deposit, Behring's Island.

age. Formerly it was abundant along the shores of Kamtchatka, the Kurile Islands, and Alaska. It was first discovered by the German naturalist, Steller, who, in company with Vitus Behring, a captain in the Russian Navy and a celebrated navigator of the northern seas, was with his vessel and crew cast away upon Behring's Island (where Behring died) in 1741. Steller's original description is preserved in the *Memoirs of the Academy of Sciences St. Petersburg.* He saw it alive during his long enforced residence on the island. In the course of forty years, 1742–1782, it appears to have been exterminated, probably for the sake of its flesh and hide, around both Behring's Island and Copper Island, to the shores of which it was, in Steller's time, limited.

Fig. 58 shows its skeleton, 19 ft. 6 in. long, now preserved in the Geological Collection at South Kensington (Glass-case N). The skeletons are found, in the islands, at a distance from the shore in old raised beaches and peat-mosses, deeply buried and thickly overgrown with grass. They are discovered by boring into the peat with an iron rod, just as timber is found in Irish peat-bogs. (See restoration, Plate XXVI.)

Steller records that when he came to Behring's Island, the Sea-cows fed in the shallows along the shore, and collected in herds like cattle. Every few minutes they raised their heads in order to get more air before descending again to browse on the thick sea-weed (probably Laminaria) surrounding the coast. With regard to their habits, they were very slow in their movements : mild and inoffensive in disposition. Their colour was dark brown, sometimes varied with spots. The skin was naked; but thick, hard, and rugged. They are said to have sometimes reached a length of thirty-five feet, when full grown. Most of their time was spent in browsing, and whilst so occupied, were not easily disturbed. Their attachment to each other was great, so that when one was harpooned, the others made great attempts to rescue it. According to Steller, they were so heavy that it required forty men with ropes to drag the body of one to land.

When, in 1743, the news of the discovery of Behring Island reached Kamtchatka, several expeditions were fitted out for the purpose of hunting the sea-cow and the various fur-bearing animals, such as the sea-otter, fur seal, and blue fox, which are found there ; and very soon many whaling vessels began to stop there to lay in a supply of sea-cow meat for food. So great was

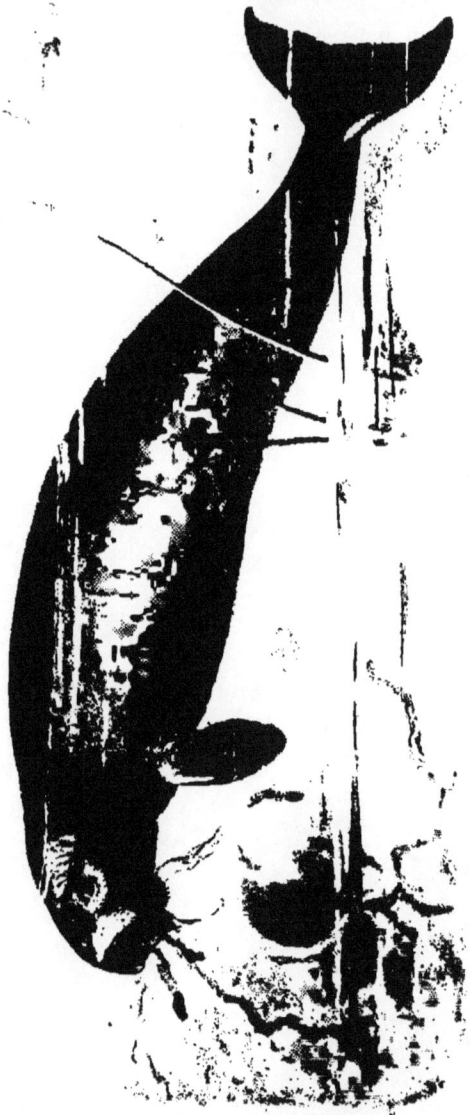

the destruction wrought by these whalers and fur-hunters that in 1754, only thirteen years after its discovery, the sea-cow had become practically exterminated. In 1768, according to the investigations of Dr. L. Stejneger of the U. S. National Museum, Washington, who has made a most careful study of the question, this large and important marine mammal became wholly extinct, the last individual ever seen alive having been killed in that year ; and the fate which overtook Rhytina so speedily has almost become that of the buffalo, and will as certainly become that of the fur seal unless it be protected.

It may interest the reader, especially if he be a traveller, to know that, besides the fine specimen of Rhytina in the Natural History Museum, already alluded to, good skeletons are possessed by the Museums of St. Petersburg, Helsingsfors (Finland), Stockholm, U. S. National Museum, Washington, as well as portions of skeletons by other museums.

The Sirenians are an ancient race, for their remains have been found in Tertiary strata, of various ages, from Eocene to Pleistocene, over the greater part of Europe—in England, Holland, Belgium, France, Germany, Austria, and Italy ; also near Cairo. In the New World, fossil Sirenians have been found in South Carolina, New Jersey, and Jamaica.

Another European species is the Halitherium, from the Miocene rocks of Hesse-Darmstadt, of which a cast may be seen in the Natural History Museum, South Kensington. Its length is 7 ft. 8 in. The teeth in this form resembled those of the Dugong.

The Rhytina was probably intermediate between the Dugong and the Manatee, judging from the casts of its brain-cavity. Its brain was very small considering the size of the animals. Altogether, as many as fourteen fossil genera and thirty species are known. Evidently, then, the old Sirenia were once a much more flourishing race. At present, they are confined exclusively to the tropical regions of the earth, and their past distribution, as revealed to the geologists, adds one more proof to the now well-established fact, that throughout most of the Tertiary era the climate of northern latitudes was very much warmer than now—in fact, sub-tropical. What cause, or causes, brought about so great a change, we cannot stay to consider here.

In conclusion, it only remains to express a hope that the reader may have been interested in our humble endeavours to describe some of the largest, most strange, and wonderful forms of life that in remote ages have found a home on this planet. And perhaps a few of our readers may be induced to add a new and never-failing interest to their lives by searching in the stony record for traces of the world's " lost creations." If so, our labour will not have been in vain,

APPENDIX I.

TABLE OF STRATIFIED ROCKS.

Periods.	Systems.	Formations.	
Quaternary. (CAINOZOIC.)	**RECENT**	Terrestrial, Alluvial, Estuarine, and Marine Beds of Historic, Iron, Bronze, and Neolithic Ages	Dominant type, Man
	PLEISTOCENE	Peat, Alluvium, Loess Valley Gravels, Brickearths Cave-deposits Raised Beaches Palæolithic Age Boulder Clay and Gravels	
Tertiary. (CAINOZOIC.)	**PLIOCENE**	Norfolk Forest-bed Series Norwich and Red Crags Coralline Crag (Diestian)	Dominant types, Birds and Mammals
	MIOCENE	Œningen Beds Freshwater, etc. Fluvio-marine Series (Oligocene)	
	EOCENE	Bagshot Beds } (Nummulitic London Tertiaries } Beds)	
SECONDARY, OR MESOZOIC.	**CRETACEOUS**	Maestricht Beds Chalk Upper Greensand Gault Lower Greensand } Neocomian Wealden }	
	JURASSIC	Purbeck Beds Portland Beds Kimmeridge Clay (Solenhofen Beds) Corallian Beds Oxford Clay Great Oolite Series Inferior Oolite Series Lias	Dominant type, Reptilia
	TRIASSIC	Rhætic Beds Keuper Muschelkalk Bunter	

TABLE OF STRATIFIED ROCKS—*Continued.*

Periods.	Systems.	Formations.	
PRIMARY, OR PALÆOZOIC.	PERMIAN or DYAS	Red Sandstone, Marl ⎱ Zech- Magnesian Limestone, etc. ⎰ stein Red Sandstone and Conglomerate 　　Rothliegende	Dominant type, Fishes
	CARBONIFEROUS {	Coal Measures and Millstone Grit Carboniferous Limestone Series	
	DEVONIAN & OLD RED SANDSTONE.	Upper Old Red Sandstone Devonian Lower Old Red Sandstone	
	SILURIAN	Ludlow Series Wenlock Series Llandovery Series May Hill Series	
	ORDOVICIAN	Bala and Caradoc Series Llandeilo Series Llanvirn Series Arenig and Skiddaw Series	
	CAMBRIAN	Tremadoc Slates Lingula Flags Menevian Series Harlech and Longmynd Series	
	EOZOIC— ARCHÆAN	Pebidian, Arvonian, and Dimetian Huronian and Laurentian	Dominant type, In-vertebrata

APPENDIX II.

MR. HENRY LEE, formerly naturalist to the Brighton Aquarium, discusses the question of " The Great Sea-Serpent" in an interesting little book, entitled *Sea Monsters Unmasked*, illustrated (1883), published as one of the Handbooks issued in connection with the International Fisheries Exhibition. He goes fully into the history of the subject, and shows how some of the appearances described may be accounted for ; but yet is inclined to think that there may exist in the sea animals of great size unknown to science, and concludes as follows :—

"This brings us face to face with the question, ' Is it, then, so impossible that there may exist some great sea creature, or creatures, with which zoologists are hitherto unacquainted, that it is necessary in every case to regard the authors of such narratives as wilfully untruthful or mistaken in their observations, if their descriptions are irreconcilable with something already known?' I, for one, am of the opinion that there is no such impossibility. Calamaries or squids of the ordinary size have, from time immemorial, been amongst the commonest and best known of marine animals in many seas ; but only a few years ago any one who expressed his belief in one formidable enough to capsize a boat or pull a man out of one was derided for his credulity, although voyagers had constantly reported that in the Indian seas they were so dreaded that the natives always carried hatchets with them in their canoes, with which to cut off the arms or tentacles of these creatures, if attacked by them. We now know that their existence is no fiction ; for individuals have been captured measuring more than fifty feet, and some are reported to have measured eighty feet in total length. As marine snakes some feet in length, and having fin-like tails adapted for swimming, abound over an extensive range, and are frequently met with far at sea, I cannot regard it as impossible that some of these also may attain to an abnormal and colossal development. Dr. Andrew Wilson, who has given much attention to this subject, is of the opinion that 'in this huge development of ordinary forms we discover the true and natural law of the production of the giant serpent of the sea.' It goes far at any rate towards accounting for its supposed appearance. I am convinced that whilst naturalists have been searching amongst the vertebrata for a solution of the problem, the great unknown, and therefore unrecognised, Calamaries, by their elongated cylindrical bodies and peculiar mode of swimming, have played the part of the sea-serpent in many a well-authenticated incident. In other cases, such as those mentioned by 'Pontoppidan' (*History of Norway*), the supposed vertical undulations of the snake seen out of water have been the burly bodies of so many porpoises swimming in line—the connecting undulations beneath the surface have been supplied by the imagination. The dorsal fins of basking sharks, as figured by Dr. Andrew Wilson, may have furnished the 'ridge of fins ;' an enormous conger is not an impossibility ; a giant turtle may have done duty, with its propelling flippers and broad back ; or a marine snake of enormous size may really have been seen. But if we accept as accurate the observations recorded (which I certainly do not in all cases, for they are full of errors and mistakes), the difficulty is not entirely met, even by this last admission, for the instances are very few in which an Ophidian proper—a true serpent—is indicated. There has seemed to be wanting an animal having a long snake-like neck, a small head, and a slender body, and propelling itself by paddles.

"The similarity of such an animal to the Plesiosaurus of old was remarkable. That curious compound reptile, which has been compared with 'a snake threaded through the body of a turtle,' is described by Dean Buckland as having 'the head of a lizard, the teeth of a crocodile, a neck of enormous length resembling the body of a serpent, the ribs of a cameleon, and the paddles of a whale.' In the number of its cervical vertebræ (about thirty-three) it surpasses that of the longest-necked bird, the swan.

"The form and probable movements of this ancient Saurian agree so markedly with some of the accounts given of 'the great sea-serpent,' that Mr. Edward Newman advanced the opinion that the closest affinities of the latter would be found to be with the Enaliosaurians, or Marine Lizards, whose fossil remains are so abundant in the Oolite and the Lias. This view has been taken by other writers, and emphatically by Mr. Gosse. Neither he nor Mr. Newman insist that 'the great unknown' must be the Plesiosaurus itself. Mr. Gosse says, ' I should not look for any species, scarcely for any genus, to be perpetuated from the Oolitic period to the present. Admitting the actual continuation of the order Enaliosauria, it would be, I think, quite in conformity with general analogy to find some salient features of several extinct forms.'

" The form and habits of the recently recognised gigantic cuttles account for so many appearances which, without knowledge of them, were inexplicable when Mr. Gosse and Mr. Newman wrote, that I think this theory is not forced upon us. Mr. Gosse well and clearly sums up the evidence as follows : 'Carefully comparing the independent narratives of English witnesses of known character and position, most of them being officers under the Crown, we have a creature possessing the following characteristics : (1) The general form of a serpent ; (2) great length, say above sixty feet ; (3) head considered to resemble that of a serpent; (4) neck from twelve to sixteen inches in diameter ; (5) appendages on the head, neck, or back, resembling a crest or mane (considerable discrepancy in details); (6) colour, dark brown or green, streaked or spotted with white ; (7) swims at surface of the water with a rapid or slow movement, the head and neck projected and elevated above the surface ; (8) progression steady and uniform, the body straight, but capable of being thrown into convolutions ; (9) spouts in the manner of a whale ; (10) like a long "nun-buoy." ' He concludes with the question, 'To which of the recognised classes of created beings can this huge rover of the ocean be referred ?'

" I reply, 'to the Cephalopoda.' There is not one of the above judiciously summarised characteristics that is not supplied by the great Calamary, and its ascertained habits and peculiar mode of locomotion.

" Only a geologist can fully appreciate how enormously the balance of probability is contrary to the supposition that any of the gigantic marine Saurians of the secondary deposits should have continued to live up to the present time. And yet I am bound to say that this does not amount to an impossibility, for the evidence against it is entirely negative. Nor is the conjecture that there may be in existence some congeners of these great reptiles inconsistent with zoological science. Dr. J. E. Gray, late of the British Museum, a strict zoologist, is cited by Mr. Gosse as having long ago expressed his opinion that some undescribed form exists which is intermediate between the tortoises and the serpents." (This is quoted by Mr. Lee in a footnote.)

" Professor Agassiz, too, is adduced by a correspondent of the *Zoologist* (p. 2395), as having said concerning the present existence of the Enaliosaurian type, that 'it would be in precise conformity with analogy that such an animal should exist in the American seas, as he had found numerous instances in which the fossil forms of the old world were represented by living types in the new.'

" On this point, Mr. Newman records in the *Zoologist* (p. 2356), an actual testimony which he considers 'in all respects the most interesting natural history fact of the present century.' He writes—

"' Captain the Hon. George Hope states that when in H.M.S. *Fly*, in the Gulf of California, the sea being perfectly calm, he saw at the bottom a large marine animal with the head and general figure of the alligator, except that the neck was much longer, and that instead of legs the creature had four large flappers, somewhat like those of turtles, the anterior pair being larger than the posterior ; the creature was distinctly visible, and all its movements could be observed with ease ; it appeared to be pursuing its prey at the bottom of the sea ; its movements were somewhat serpentine, and an appearance of annulations, or ring-like divisions of the body, was distinctly perceptible. Captain Hope made this relation in company, and as a matter of conversation. When I heard it from the gentleman to whom it was narrated, I inquired whether Captain Hope was acquainted with those remarkable fossil animals, Ichthyosauri and Plesiosauri, the supposed forms of which so nearly corresponded with what he describes as having seen alive, and I cannot find that he had heard of them ; the alligator being the only animal he mentioned as bearing a partial similarity to the creature in question.'

" Unfortunately, the estimated dimensions of this creature are not given.

" That negative evidence alone is an unsafe basis for argument against the existence of unknown animals, the following illustrations will show :—

" During the deep-sea dredgings of H.M.S. *Lightning*, *Porcupine*, and *Challenger*, many new species of mollusca and others, which had been supposed to have been extinct ever since the Chalk, were brought to light ; and by the deep-sea trawlings of the last-mentioned ship there have been brought up from great depths fishes of unknown species, and which could not exist near the surface, owing to the distention and rupture of their air-bladder when removed from the pressure of deep water.

" Mr. Gosse mentions that the ship in which he made the voyage to Jamaica was surrounded in the North Atlantic, for seventeen continuous hours, by a troop of whales of large size, of an undescribed species, which on no other occasion has fallen under scientific observation. Unique specimens of other Cetaceans are also recorded. "We have evidence, to which attention has been directed by Mr. A. D. Bartlett, that 'even on land there exists at least one of the largest mammals, probably in thousands, of which only one individual has been brought to notice, namely, the hairy-eared, two-horned rhinoceros (*R. Lasiotis*), now in the Zoological Gardens, London. It was captured in 1868, at Chittagong, in India, where for years collectors and naturalists have worked and published lists of the animals met with, and yet no knowledge of this great beast was ever before obtained, nor is there any portion of one in any museum. It remains unique.

" I have arrived at the following conclusions : 1. That without straining resemblances, or casting a doubt upon narratives not proved to be erroneous, the various appearances of the supposed 'great sea-serpent' may now be nearly all accounted for by the forms and habits of known animals ; especially if we admit, as proposed by Dr. Andrew Wilson, that some of them, including the marine snakes, may, like the cuttles, attain to extraordinary size. 2. That to assume that naturalists have perfect cognisance of every existing marine animal of large size, would be quite unwarrantable. It appears to me more than probable that many marine animals, unknown to science, and some of them of gigantic size, may have their ordinary habitat in the sea, and only occasionally come to the surface ; and I think it not impossible that amongst them may be marine snakes of greater dimensions than we are aware of, and even a creature having close affinities with the old sea-reptiles whose fossil skeletons tell of their magnitude and abundance in past ages.

" It is most desirable that every supposed appearance of 'the Great Sea-Serpent' shall be faithfully noted and described ; and I hope that no truthful observer will be deterred from reporting such an occurrence by fear of the disbelief of naturalists or the ridicule of witlings."

S

APPENDIX III.

1. FROM RIVER VALLEYS AND ALLUVIAL DEPOSITS.

ENGLAND.

Cornwall and Devonshire.—None.

Somersetshire.—Hinton, Larkhall, Hartlip, St. Audries, Weston-super-Mare, Chedzoy, Freshford.

Gloucestershire.—Gloucester, Barnwood, Beckford, Stroud, Tewkesbury.

Dorsetshire.—Bridport, Portland Fissure.

Hampshire.—Gale Bay, Newton.

Wiltshire.—Christian Malford, Fisherton, Milford Hill, near Salisbury.

Berkshire.—Maidenhead, Taplow, Reading, Hurley Bottom.

Oxfordshire.—Yarnton, Bed of the Cherwell, City of Oxford, Wytham, Culham.

Essex.—Lexden, Orford, Hedingham, Lamarsh, Isle of Dogs, Walton-on-the-Naze, Ilford (the finest specimen, see p. 187), Wenden, Harwich, Colchester, Ballingdon, Walthamstow.

Hertfordshire.—Camp's Hill.

Sussex.—Bracklesham Bay, Brighton, Lewes, Valley of Arun, Pagham.

Suffolk.—Ipswich, Hoxne.

Norfolk.—Bacton, Cromer, Yarmouth.

Cambridge.—Barrington, Barnwell, Chesterton, Great Shelford, Barton, Westwick Hall.

[1] From Mr. Leith Adams's Monograph on *British Fossil Elephants.* Palæontographical Society, London. 1877.

Huntingdonshire.—Huntingdon, St. Neots.

Bedfordshire.—Leighton Buzzard.

Middlesex.—At London, under various streets, etc., viz., St. James's Square, Pall Mall, Kensington, Battersea, Hammersmith, and, recently (1892), in Endsleigh Street. Turnham Green. In the bed of the Thames at Millbank, Brentford, Kew, Acton, Clapton, etc. Kingsland..

Surrey.—Wallington, Tooting, Peckham, Dorking, Peasemarsh, near Guildford.

Kent.—Crayford, Erith, Dartford, Aylesford, Hartlip, Otterham, Isle of Sheppey, Broughton Fissure, Medway, Sittingbourne, Newington, Green Street Green, Bromley, Whitstable.

Buckinghamshire.—Fenny Stratford.

Northamptonshire.—Oundle, Kettering, Northampton.

Warwickshire.—Rugby, Wellesborne, Lawford, Bromwich Hill, Halston, Newnam.

Worcestershire.—Stour Valley, Droitwich, Banks of Avon, Fladbury, Malvern.

Leicestershire.—Kirby Park.

Staffordshire.—Copen Hall, Trentham.

Cheshire.—Northwich.

Lincolnshire.—Spalding.

Yorkshire.—Whitby, Aldborough, Gristhorpe Bay, Harswell, Leeds, Bielbecks, Brandsburton, Middleton, Overton, Alnwick, Hornsea.

Herefordshire.—Kingsland.

SCOTLAND.

Ayrshire.—Kilmaurs.

Between Edinburgh and Falkirk.

Chapel Hall in Lanarkshire, and Bishopbriggs.

At Clifton Hall.

IRELAND.

Cavan.—Belturbet.

Antrim.—Corncastle.

Waterford.—Near Whitechurch (but somewhat doubtful).

2. FROM CAVERNS.

Devonshire.—Kent's Cavern, Oreston, Beach Cave, Brixham.

Somerset.—Hutton Cave, and a cave near Wells, Wookey Hole, Bleadon Cave, Box Hill, near Bath, Durdham Down, Sandford Hill.

Kent.—In Boughton Cave, near Maidstone.

Nottinghamshire.—In Church Hole.

Derbyshire.—In Cresswell Crags, Robin Hood Cave, Church Hole.

Glamorganshire.—In Long Hole, Spritsail Tor, Paviland.

Caermarthen.—In Coygan Cave.

Waterford.—In Shandon Cave.

APPENDIX IV.

LITERATURE.

1. POPULAR WORKS.

The Story of the Earth and Man. By Sir Wm. Dawson.
The Mammoth and the Flood. By Sir Henry Howorth.
Works by Doctor Gideon A. Mantell :—
  Medals of Creation.
  Wonders of Geology.
  Petrifactions and their Teaching.
Phases of Animal Life. By R. Lydekker.
Science for All. 5 vols. (Chapters on Extinct Animals.)
Our Earth and its Story, vol. ii.
Winners in Life's Race. By Arabella Buckley.
The Autobiography of the Earth. By Rev. H. N. Hutchinson.
Sea Monsters Unmasked. By H. Lee.

2. WORKS OF REFERENCE.

A Manual of Palæontology. 2 vols. By Prof. Alleyne Nicholson,
and R. Lydekker.
The Life-History of the Earth. By Prof. Alleyne Nicholson.
Origin of Species. By C. Darwin. Also *The Journal of Researches*.
The Old Red Sandstone. By Hugh Miller.
Sketch Book of Popular Geology. By Hugh Miller.
Early Man in Britain. By Prof. Boyd Dawkins.
The English Encyclopedia. (The 2 vols. on Natural History contain
much information on extinct animals.)
The Encyclopedia Britannica. Ninth Edition.
Memoirs of the Ichthyosauri and Plesiosauri. By Thos. Hawkins.

Phillips's *Manual of Geology.*   New Edition, by Prof. H. G. Seeley and R. Etheridge.

The Book of the Great Sea-Dragons.   By Thos. Hawkins.

The Geographical and Geological Distribution of Animals.   By A. Heilprin.

Prehistoric Europe.   By Prof. James Geikie.

Palæontological Memoirs.   By Hugh Falconer, M.D.

Mammals, Living and Extinct.   By Prof. Flower and R. Lydekker.

British Fossil Mammals and Birds.   By Sir R. Owen.

A Manual of Palæontology.   By Sir R. Owen.

A Catalogue of British Fossil Vertebrata.   By A. S. Woodward and C. D. Sherborn.

3. MONOGRAPHS.

The Dinocerata.   By Prof. O. C. Marsh.   *United States Geological Survey*, vol. x.   Washington, 1884.

The Odontornithes, a Monograph on the Extinct Toothed Birds of North America.   By Prof. O. C. Marsh.   New Haven, Connecticut, 1880.

The Vertebrata of the Tertiary Formations.   By Prof. E. D. Cope. Washington, 1883.

The Vertebrata of the Cretaceous Formations of the West.   By Prof. E. D. Cope.   Washington, 1875.

Contributions to the Extinct Vertebrate Fauna of the Western Territories.   By Joseph Leidy.   Washington, 1873.

(The last three are in the reports of the *United States Geological Survey of the Territories.*)

The British Merostomata (Palæontographical Society).   By Dr. Henry Woodward, F.R.S.

MONOGRAPHS BY SIR RICHARD OWEN.

A History of British Fossil Reptiles.   4 vols.   (Cassell.)   (Most of which has been previously published in the *Monographs of the Palæontographical Society.*)

On the Megatherium, or Giant Ground Sloth of America.   London, 1860.

On the Mylodon.   London, 1842.

On the Extinct Wingless Birds of New Zealand.   London, 1878. Reprinted from *The Transactions of the Zoological Society.*

4. JOURNALS.

The student should consult the numerous papers by Prof. Marsh in *The American Journal of Science*; and by Prof. Cope in *The American Naturalist.* Many of Prof. Marsh's papers have also appeared in *The Geological Magazine* and in *Nature.* The two latter journals contain many other valuable papers (and reviews of Monographs, etc.), too numerous to be separately mentioned. Some are referred to in the text. *The Quarterly Journal of the Geological Society* contains many papers on Extinct Animals. See also papers in *Natural Science* and *Knowledge.*

APPENDIX V.

IT was unfortunate that news of the highly interesting discovery at Würtemberg came too late for our artist to make a new drawing

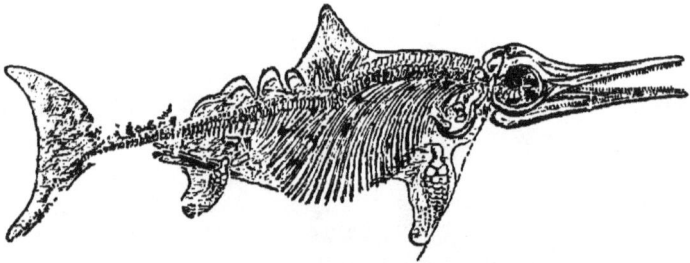

FIG. 59.—*Ichthyosaurus tenuirostris*, from Würtemberg.

for our first edition, to show the dorsal fin and large tail-fin, etc., described by Dr. Fraas.[1] This has now been done, as shown in

---

[1] Ueber einen neuen Fund von *Ichthyosaurus* in Würtemberg. *Neues Jahrbuch f. Mineralogie*, 1892, vol. ii. pp. 87-90. The same author has published a valuable monograph, with beautiful plates, entitled *Die Ichthysaurier der Süddentschen Trias- und Jura-Ablagerungen*. 4to. Tübingen, 1891.

Plate II. By the courtesy of the proprietors of *Natural Science*, we are enabled to reproduce two drawings (Fig. 59) from the September number, illustrating a paper by Mr. Lydekker, in which he gives a *résumé* of the latest intelligence with regard to Ichthyosaurian reptiles.

In the present year (1892) there has been discovered in the Lias of Würtemberg the skeleton of an Ichthyosaur, in which the outline of the fleshy parts is completely preserved (see lower figure). The reader will see from the figure that the tail-fin is very large, and the backbone appears to run into the lower lobe. Such a tail-fin as this impression indicates must have resembled that of the shark's, only it is wider; but the shark's backbone runs into the *upper* lobe. Sir Richard Owen long ago foretold the existence of this appendage, and the discovery, coming now (when his life is despaired of), adds one more tribute to his genius. Behind the triangular fin on the back comes a row of horny excrescences reminding us of those of the crested newt.

As Dr. Fraas remarks, this discovery shows how closely analogous Ichthyosaurs were in form to fishes, and further justifies the title of "fish-lizards." He considers that they did *not* visit the shore. The reader will find much valuable matter in Mr. Lydekker's paper, above referred to. The following extract refers to the question of their reproduction : " It has long been known that certain large skeletons of Ichthyosaurs from the Upper Lias of Holzmaden, in Würtemberg, and elsewhere, are found with the skeletons of one or more much smaller individuals enclosed partly or entirely within the cavity of the ribs [a specimen is figured]. Of such skeletons there are four in the museum at Stuttgart, two in that of Tübingen, one at Munich, and others in Gent and Paris. Of these, two in Stuttgart, as well as the two in Tübingen, contain but a single young skeleton, while one of those at Stuttgart has four, the Munich specimen five, and the remaining Stuttgart example upwards of seven young. Some of these young and, presumably, fœtal Ichthyosaurs have the head turned towards the tail of the parent, while in others it is directed the other way. That these young have not been swallowed by the larger individuals within whose ribs they are found is pretty evident from several considerations. In the first place, their skeletons are always perfect. Then they never exceed one particular size, and always belong to the same species as the parent. Moreover, it would appear to be a physical impossibility for one Ichthyosaur of the size of the Stuttgart specimen to have had seven smaller ones of such dimensions in its stomach at one and the same time.   We may accordingly take it for granted that these imprisoned skeletons were those of fœtuses.   It is, however, very remarkable, that, so far as we are aware, all the skeletons with fœtuses belong to one single species ; thus suggesting that this particular species was alone viviparous."

It is to be hoped that further discoveries will be made, such as may finally settle this question.   One would have expected that in some cases the young ones, if fœtal, would be imperfectly developed.

INDEX.

A

ÆPYORNIS. Vid. Moa-bird.
Agassiz, 27
"Age of Reptiles," 63, 107; "Age of Mammals," 147
Air, action of, 10
Allosaurus, 83
Ancients, ideas of the, 35, 61, 155, 195, 199
Apatosaurus, 70
Aqueous rocks, 14
Arbroath paving-stone, 26
Armadillo. Vid. Glyptodon.
Articulata, 25
Atlantosaurus, 70

B

Backbone of fishes, 49
"Bad Lands" of Wyoming, 157
Baker, Sir Samuel, on Crocodiles, 48 ; on Elephants, 215
Basalt, 14
Berossus, the Chaldæan, 34
Birds, fossilisation of, 19; ancestry of, 63, 109. Vid. Hesperornis, Moa.
Blackie, Prof. J. S., on Ichthyosaurus, 37
"Breaks," 21, 147
Brontops, 160
Brontosaurus, 66; vertebræ of, 68; habits of, 69
Buckland, Dean, 37, 46, 53, 73, 75–77, 124, 126, 180
Buffon, 5, 223

C

Cautley, Captain, 162
Cave-earth, 10

Ceratosaurus, 84
Cetiosaurus, 73, 74
Challenger, H.M.S., 20
Chinese legends of Mammoth, 199
Clidastes, 144, 145
Climate, of Lias period, 51 ; of Eocene period, 159 ; of Tertiary era, 163
Collini, 123
Compsognathus, 86
Conybeare, Rev., on Plesiosaurus, 52, 58 ; on Sea-serpents, 135
Cope, Prof. E. D., on Sea-serpents, 139, 141, 143 ; on Eocene wingless bird, 237
Correlation, law of, 6, 43, 54, 88, 161
Crustaceans, 24
Cuvier, 2, 5, 7, 63, 73, 76; on Ichthyosaurus, 36 ; on Plesiosaurus, 53 ; on Iguanodon teeth, 90, 91 ; on Pterodactyls, 121, 122, 126 ; on Mosasaurus, 135, 136; on Tertiary animals, 148; on Megatherium, 179; on Mammoth, 193, 212 ; on Mastodon, 217

D

Darwin, Charles, 20 ; on extinct Sloths, 181
Dawkins, Prof. Boyd, 10 ; on Mammoth, 213
De la Beche, Sir Henry, 37, 52
Denudation, 21
Dimorphodon, 124
Dinocerata, 149 ; skull and limbs of, 150 ; where found, 155
Dinornis. Vid. Moa-bird.
Dinosaurs, chaps. v., vi., vii. ; anatomy of, 64 ; geographical range of, 75 ;

Recently Published, by the same Author. Price 5s.

THE
STORY OF THE HILLS.

A POPULAR ACCOUNT
OF MOUNTAINS AND HOW THEY WERE MADE.

BY THE

REV. H. N. HUTCHINSON, B.A., F.G.S.,

AUTHOR OF "THE AUTOBIOGRAPHY OF THE EARTH."

OPINIONS OF THE PRESS.

"This work belongs to that useful class whose intention is to arouse interest in the works of nature, and quicken the faculty of observation."
Manchester Guardian.

"It tells in the pleasantest way the first things that geologists learn and teach crabbedly about the heaving up of hills, the wearing of them down by the weather, the breaking out of volcanoes, and kindred matters."—*Scotsman.*

"The author is a man of wide geological and physiographical reading, possessed of the gift of clearly interpreting the writers he reads, and of reproducing their facts and conclusions in easily understood and even attractive language."—*Science Gossip.*

"It will be read with pleasure and profit by the tourist who likes to know just enough about the sundry points of interest connected with the scene of his wanderings to make the enjoyment of his outing intelligent."—*Nature.*

"Mr. Hutchinson's book deals with the slow moulding of mountain forms by streams and by weathering, and with the forces by which mountains have been upheaved, and will double the pleasure of a mountain trip. It is of a handy and portable size, and is illustrated with several excellent reproductions of photographs by the late Mr. W. Donkin."—*Knowledge.*

"A charmingly written and beautifully illustrated account of the making of the mountains. An admirable gift book."—*Yorkshire Post.*

"This is a popular and well illustrated account of mountains and how they were made. The illustrations are especially excellent, being reproductions of photographs taken by the late Mr. W. Donkin, Messrs. Walentine and Sons (Dundee), and Mr. Wilson (Aberdeen). Mr. Hutchinson writes interestingly, and evidently knows geology and physiography."—*Journal of Education.*

SEELEY AND CO., LIMITED, ESSEX STREET, STRAND.

[OVER.

Recently Published, by the same Author.

THE
AUTOBIOGRAPHY OF THE EARTH.

A POPULAR ACCOUNT OF GEOLOGICAL HISTORY.

BY THE

REV. H. N. HUTCHINSON, B.A., F.G.S.

Crown 8vo, cloth, with 27 Illustrations, price 7s. 6d.

CONTENTS.—1. Cloud-land, or Nebular Beginnings—2. The Key to Geology—3. An Archaic Era—4. Cambrian Slates—5. The Slates and Ashes of Siluria—6. The Old Red Sandstone—7. The Mountain Limestone—8. Forests of the Coal-period—9. A Great Interval—10. The Cheshire Sandstones—11. New Phases of Life—12. Bath Oolites—13. An Age of Reptiles—14. The Chalk Downs—15. The New Era—16. The Ice-Age and Advent of Man.

SOME OPINIONS OF THE PRESS.

"His sketch of historic geology has a genuine continuity. It is so written as to be understanded of plain people, and is illustrated by some very good woodcuts and diagrams.' —*Saturday Review.*

"This most interesting book."—*Spectator.*

"A delightfully written and thoroughly accurate popular work on geology, well calculated to engage the interest of readers in the fascinating study of the Stony Science."—*Science Gossip.*

"In this work the Rev. H. N. Hutchinson produces a popular account of geological history, and explains the principles and methods by which that history has been read. He endeavours to interpret the past by the light of the present, first acquiring a knowledge, by direct observation and self-instruction, of the chief operations now taking place on the earth's surface, and then employing this knowledge to ascertain the meaning of the record of stratified rocks. This principle of 'uniformity' knocked the old teaching of catastrophism on the head. The author is accurate in all his details, yet his subject is touched into something not at all unlike romance. The illustrations are good."—*National Observer.*

LONDON: EDWARD STANFORD, 26 & 27, COCKSPUR STREET, S.W.

11, HENRIETTA STREET, W.C.,
February, 1893.

BOOKS

PUBLISHED BY

CHAPMAN & HALL, Ld.

(A SELECTION.)

Afialo (F. G.) and Surgeon-General C. T. Paske.
THE SEA AND THE ROD. With Illustrations. Crown 8vo, 4s. 6d.

Anderson (Andrew A.).
A ROMANCE OF N'SHABÈ: Being a Record of Startling Adventures in South Central Africa. With Illustrations. Crown 8vo, 5s.

TWENTY-FIVE YEARS IN A WAGGON IN THE GOLD RE-GIONS OF AFRICA. With Illustrations and Map. Second Edition. Demy 8vo, 12s.

Anderson (Captain Lindsay).
THE STORY OF ALLAN GORDON. With Illustrations. Crown 8vo. 5s.

AMONG TYPHOONS AND PI-RATE CRAFT. With Illustrations. Crown 8vo, 5s.

A CRUISE IN AN OPIUM CLIP-PER. With Illustrations. Crown 8vo, 6s.

Bailey (J. B.).
FROM SINNER TO SAINT: or, Character Transformations; being a few Biographical Sketches of His-toric Individuals whose Moral Lives underwent a Remarkable Change. Crown 8vo, 6s.

Baker (W. L.), A.M.I.C.E.
THE BEAM; or, Technical Elements of Girder Construction. Crown 8vo, 4s.

Bell (James, Ph.D., etc.), *Principal of the Somerset House Laboratory.*
THE CHEMISTRY OF FOODS. With Microscopic Illustrations.

Part I. TEA, COFFEE, COCOA, SUGAR, etc. Large crown 8vo, 2s. 6d.

Part II. MILK, BUTTER, CHEESE, CEREALS, PREPARED STARCHES, etc. Large crown 8vo, 3s.

Bentley (H. Cumberland).
SONGS AND VERSES. Illustrated by FINCH MASON, and dedicated to J. G. WHYTE MELVILLE. Crown 8vo, 4s.

Birdwood (Sir George C. M.), C.S.I.
THE INDUSTRIAL ARTS OF INDIA. With Map and 174 Illustra-tions. New Edition. Demy 8vo, 14s.

Blatherwick (C.).
IN THE SHADE OF SCHIEHAL-LION. With 8 Illustrations. One Volume. Crown 8vo, 1s.

CYNTHIA. With 4 Illustrations. One Volume. Crown 8vo, 1s.

Brackenbury (Col. C. B.).
FREDERICK THE GREAT. With Maps and Portrait. Large crown 8vo, 4s.

Bradley (Thomas), *of the Royal Military Academy, Woolwich.*
ELEMENTS OF GEOMETRICAL DRAWING. In Two Parts, with 60 Plates. Oblong folio, half bound, each Part 16s.

Bridgman (F. A.).

WINTERS IN ALGERIA. With 62 Illustrations. Royal 8vo, 10s. 6d.

BRITISH ARMY, THE. By the Author of "Greater Britain," "The Present Position of European Politics," etc. Demy 8vo, 12s.

Bromley-Davenport (the late W.), M.P.

SPORT: Fox Hunting, Salmon Fishing, Covert Shooting, Deer Stalking. With numerous Illustrations by GENERAL CREALOCKE, C.B. Cheap Edition. Post 8vo, 3s. 6d.

Buckland (Frank).

LOG-BOOK OF A FISHERMAN AND ZOOLOGIST. With numerous Illustrations. Sixth Thousand. Crown 8vo, 3s. 6d.

Burgess (Edward).

ENGLISH AND AMERICAN YACHTS. Illustrated with 50 beautiful Photogravure Engravings. Oblong folio, 42s.

Carstensen (A. Riis).

TWO SUMMERS IN GREENLAND: An Artist's Adventures among Ice and Islands in Fjords and Mountains. With numerous Illustrations by the Author. Demy 8vo, 14s.

CHARLOTTE ELIZABETH, LIFE AND LETTERS OF, Princess Palatine and Mother of Philippe d'Orléans, Regent of France, 1652-1722. With Portraits. Demy 8vo, 10s. 6d.

Charnay (Désiré).

THE ANCIENT CITIES OF THE NEW WORLD. Being Travels and Explorations in Mexico and Central America, 1657-1882. With upwards of 200 Illustrations. Super Royal 8vo, 31s. 6d.

CHRIST THAT IS TO BE, THE: A Latter-day Romance. Third Edition. Demy 8vo, 3s. 6d.

Church (Professor A. H.), M.A. Oxon.

FOOD GRAINS OF INDIA. With numerous Woodcuts. Small 4to, 6s.

ENGLISH PORCELAIN. A Handbook to the China made in England during the Eighteenth Century, as illustrated by Specimens chiefly in the National Collection. With numerous Woodcuts. Large crown 8vo, 3s.

ENGLISH EARTHENWARE. A Handbook to the Wares made in England during the Seventeenth and Eighteenth Centuries, as illustrated by Specimens in the National Collections. With numerous Woodcuts. Large crown 8vo, 3s.

PLAIN WORDS ABOUT WATER. Illustrated. Crown 8vo, sewed, 6d.

FOOD: Some Account of its Sources, Constituents, and Uses. A New and Revised Edition. Twelfth Thousand. Large crown 8vo, cloth, 3s.

PRECIOUS STONES: considered in their Scientific and Artistic Relations. With a Coloured Plate and Woodcuts. Second Edition. Large crown 8vo, 2s. 6d.

COBDEN, RICHARD, LIFE OF. By the RIGHT HON. JOHN MORLEY, M.P. With Portrait. New Edition. Crown 8vo, 7s. 6d.
Popular Edition, with Portrait, 4to, sewed, 1s.; cloth, 2s.

Collier (The Hon. Margaret), MADAME GALETTI DI CADILLIAC.

RACHEL AND MAURICE, and OTHER TALES. Crown 8vo, 3s. 6d.

Collins (Wilkie) and Dickens (Charles).

THE LAZY TOUR OF TWO IDLE APPRENTICES; NO THOROUGHFARE; THE PERILS OF CERTAIN ENGLISH PRISONERS. With 8 Illustrations. Crown 8vo, 5s.

₄ These Stories are now reprinted for the first time complete.

Cookery.

DINNERS IN MINIATURE. By MRS. EARL. Crown 8vo, 2s. 6d.

HILDA'S "WHERE IS IT?" OF RECIPES. Containing many old CAPE, INDIAN, and MALAY DISHES and PRESERVES ; and a Collection of Home Remedies in Case of Sickness. By H. J. DUCKITT. Fourth Thousand. Crown 8vo, 4s. 6d.

THE PYTCHLEY BOOK OF RE-FINED COOKERY AND BILLS OF FARE. By MAJOR L——. Fourth Edition. Large crown 8vo, 8s.

BREAKFASTS, LUNCHEONS, AND BALL SUPPERS. By MAJOR L——. Crown 8vo, 4s.

OFFICIAL HANDBOOK OF THE NATIONAL TRAINING SCHOOL FOR COOKERY. Containing Lessons on Cookery; forming the Course of Instruction in the School. Compiled by "R. O. C." Twenty-first Thousand. Large crown 8vo, 6s.

BREAKFAST AND SAVOURY DISHES. By "R. O. C." Ninth Thousand. Crown 8vo, 1s.

THE ROYAL CONFECTIONER, English and Foreign. A Practical Treatise. By C. E. FRANCATELLI. With numerous Illustrations. Sixth Thousand. Crown 8vo, 5s.

Cooper-King (Lt.-Col.).

GEORGE WASHINGTON. Large crown 8vo. With Portrait and Maps. [In the Press.

Couperus (Louis).

ELINE VERE. Translated by J. T. GREIN. Crown 8vo, 5s.

Courtney (W. L.), M.A., LL.D., of New College, Oxford.

STUDIES AT LEISURE. Crown 8vo, 6s.

STUDIES NEW AND OLD. Crown 8vo, 6s.

CONSTRUCTIVE ETHICS: A Review of Modern Philosophy and its Three Stages of Interpretation, Criticism, and Reconstruction. Demy 8vo, 12s.

Craik (George Lillie).

ENGLISH OF SHAKESPEARE. Illustrated in a Philological Commentary on "Julius Cæsar." Eighth Edition. Post 8vo, cloth, 5s.

OUTLINES OF THE HISTORY OF THE ENGLISH LANGUAGE. Eleventh Edition. Post 8vo, cloth, 2s. 6d.

Crawfurd (Oswald).

ROUND THE CALENDAR IN PORTUGAL. With numerous Illustrations. Royal 8vo, 18s.

BEYOND THE SEAS; Being the surprising Adventures and ingenious Opinions of Ralph, Lord St. Keyne, told by his kinsman, Humphrey St. Keyne. Second Edition. Crown 8vo, 3s. 6d.

Cripps (Wilfred Joseph), M.A., F.S.A.

COLLEGE AND CORPORATION PLATE. A Handbook for the Reproduction of Silver Plate. With numerous Illustrations. Large crown 8vo, cloth, 2s. 6d.

Curzon (Louis Henry).

A MIRROR OF THE TURF; or, The Machinery of Horse-racing Revealed; showing the Sport of Kings as it is To-day. Crown 8vo, 8s.

Dairy Farming.

DAIRY FARMING. To which is added a Description of the Chief Continental Systems. With numerous Illustrations. By JAMES LONG. Crown 8vo, 9s.

DAIRY FARMING, MANAGEMENT OF COWS, etc. By ARTHUR ROLAND. Edited by WILLIAM ABLETT. Crown 8vo, 5s.

Day (William).

THE RACEHORSE IN TRAINING, with Hints on Racing and Racing Reform, to which is added a Chapter on Shoeing. Seventh Edition. Demy 8vo, 9s.

De Bovet (Madame).
THREE MONTHS' TOUR IN IRELAND. Translated and Condensed by MRS. ARTHUR WALTER. With Illustrations. Crown 8vo, 6s.

De Champeaux (Alfred).
TAPESTRY. With numerous Woodcuts. Cloth, 2s. 6d.

De Falloux (The Count).
MEMOIRS OF A ROYALIST. Edited by C. B. PITMAN. 2 vols. With Portraits. Demy 8vo, 32s.

Delille (Edward).
SOME MODERN FRENCH WRITERS. Crown 8vo. [*In the Press.*

DE LISLE (MEMOIR OF LIEUTENANT RUDOLPH), R.N., of the Naval Brigade. By the Rev. H. N. OXENHAM, M.A. Third Edition. Crown 8vo, 7s. 6d.

De Mandat-Grancey (Baron E.).
PADDY AT HOME; OR, IRELAND AND THE IRISH AT THE PRESENT TIME, AS SEEN BY A FRENCHMAN. Fifth Edition. Crown 8vo, 1s.; in cloth, 1s. 6d.

De Stael (Madame).
MADAME DE STAEL: Her Friends and Her Influence in Politics and Literature. By LADY BLENNERHASSETT. Translated from the German by J. E. GORDON CUMMING. With a Portrait. 3 vols. Demy 8vo, 36s.

De Windt (H.).
SIBERIA AS IT IS. With an Introduction by MADAME OLGA NOVIKOFF. With numerous Illustrations. Demy 8vo, 18s.

A RIDE TO INDIA ACROSS PERSIA AND BELUCHISTAN. With numerous Illustrations and Map. Demy 8vo, 16s.

FROM PEKIN TO CALAIS BY LAND. With numerous Illustrations. New and Cheap Edition. Demy 8vo, 7s. 6d.

Dickens (Mary A.).
CROSS CURRENTS: a Novel. A New Edition in One Volume. Crown 8vo, 3s. 6d.

Dilke (Lady).
ART IN THE MODERN STATE. With Facsimile. Demy 8vo, 9s.

Dixon (Charles).
THE GAME BIRDS AND WILD FOWL OF THE BRITISH ISLANDS. Being a Handbook for the Naturalist and Sportsman. With Illustrations by A. T. ELWES. Demy 8vo, 18s.

THE MIGRATION OF BIRDS. An Attempt to reduce Avian Season Flight to Law. Crown 8vo, 6s.

THE BIRDS OF OUR RAMBLES: A Companion for the Country. With Illustrations by A. T. ELWES. Large crown 8vo, 7s. 6d.

IDLE HOURS WITH NATURE. With Frontispiece. Crown 8vo, 6s.

ANNALS OF BIRD LIFE: A Year-Book of British Ornithology. With Illustrations. Crown 8vo, 7s. 6d.

Douglas (John).
SKETCH OF THE FIRST PRINCIPLES OF PHYSIOGRAPHY. With Maps and numerous Illustrations. Crown 8vo, 6s.

Ducoudray (Gustave).
THE HISTORY OF ANCIENT CIVILISATION. A Handbook based upon M. Gustave Ducoudray's "Histoire Sommaire de la Civilisation." Edited by the Rev. J. VERSCHOYLE, M.A. With Illustrations. Large crown 8vo, 6s.

THE HISTORY OF MODERN CIVILISATION. With Illustrations. Large crown 8vo, 9s.

Dyce (William), R.A.
DRAWING-BOOK OF THE GOVERNMENT SCHOOL OF DESIGN. Fifty selected Plates. Folio, sewed, 5s.; mounted, 18s.

Dyce (William), R.A.—*continued.*
ELEMENTARY OUTLINES OF ORNAMENT. Plates I. to XXII., containing 97 Examples, adapted for Practice of Standards I. to IV. Small folio, sewed, 2s. 6d.

Ellis (A. B.), Colonel 1st West India Regiment.
HISTORY OF THE GOLD COAST OF WEST AFRICA. Demy 8vo.
[*In the Press.*
THE EWE-SPEAKING PEOPLE OF THE SLAVE COAST OF WEST AFRICA. With Map. Demy 8vo, 10s. 6d.
THE TSHI-SPEAKING PEOPLES OF THE GOLD COAST OF WEST AFRICA; Their Religion, Manners, Customs, Laws, Language, etc. With Map. Demy 8vo, 10s. 6d.
SOUTH AFRICAN SKETCHES. Crown 8vo, 6s.
THE HISTORY OF THE FIRST WEST INDIA REGIMENT. Demy 8vo, 18s.
THE LAND OF FETISH. Demy 8vo, 12s.

Engel (Carl).
MUSICAL INSTRUMENTS. With numerous Woodcuts. Large crown 8vo, cloth, 2s. 6d.
ENGLISHMAN IN PARIS: NOTES AND RECOLLECTIONS. 2 vols. Seventh Thousand. Demy 8vo, 18s.
Vol. I.—REIGN OF LOUIS PHILIPPE.
Vol. II.—THE EMPIRE.

Escott (T. H. S.).
POLITICS AND LETTERS. Demy 8vo, 9s.
ENGLAND: ITS PEOPLE, POLITY, AND PURSUITS. New and Revised Edition. Demy 8vo, 3s. 6d.
EUROPEAN POLITICS, THE PRESENT POSITION OF. By the Author of "Greater Britain." Demy 8vo, 12s.

Fane (Violet).
AUTUMN SONGS. Crown 8vo, 6s.
THE STORY OF HELEN DAVENANT. Crown 8vo, 3s. 6d.
QUEEN OF THE FAIRIES (A Village Story), and other Poems. Crown 8vo, 6s.
ANTHONY BABINGTON: A Drama. Crown 8vo, 6s.

Fitzgerald (Percy), F.S.A.
THE HISTORY OF PICKWICK. An Account of its Characters, Localities, Allusions, and Illustrations. With a Bibliography. Demy 8vo, 8s.

Fleming (George), F.R.C.S.
ANIMAL PLAGUES: THEIR HISTORY, NATURE, AND PREVENTION. 8vo, cloth, 15s.
PRACTICAL HORSE-SHOEING. With 37 Illustrations. Fifth Edition, enlarged. 8vo, sewed, 2s.
RABIES AND HYDROPHOBIA: THEIR HISTORY, NATURE, CAUSES, SYMPTOMS, AND PREVENTION. With 8 Illustrations. 8vo, cloth, 15s.
FORSTER, THE LIFE OF THE RIGHT HON. W. E. By T. WEMYSS REID. With Portraits. Fifth Edition, in one volume, with new Portrait. Demy 8vo, 10s. 6d.

Forsyth (Captain).
THE HIGHLANDS OF CENTRAL INDIA: Notes on their Forests and Wild Tribes, Natural History and Sports. With Map and Coloured Illustrations. A New Edition. Demy 8vo, 12s.

Fortnum (C. D. E.), F.S.A.
MAIOLICA. With numerous Woodcuts. Large crown 8vo, cloth, 2s. 6d.
BRONZES. With numerous Woodcuts. Large crown 8vo, cloth, 2s. 6d.

Franks (A. W.).
JAPANESE POTTERY. Being a Native Report, with an Introduction. With Illustrations and Marks. Large crown 8vo, 2s. 6d.

Gardner (J. Starkie).
A HISTORY OF IRON-WORKING AS AN ART. With numerous Illustrations. Large crown 8vo.
[In the Press.

Gasnault (Paul) and Garnier (Ed.).
FRENCH POTTERY. With Illustrations and Marks. Large crown 8vo, 3s.

Gleichen (Count), *Grenadier Guards.*
WITH THE CAMEL CORPS UP THE NILE. With numerous Sketches by the Author. Third Edition. Large crown 8vo, 9s.

Gower (A. R.), *Royal School of Mines.*
PRACTICAL METALLURGY. With Illustrations. Crown 8vo, 3s.

Griffiths (Major Arthur).
SECRETS OF THE PRISON HOUSE. *[In the Press.*
FRENCH REVOLUTIONARY GENERALS. Large crown 8vo, 6s.
CHRONICLES OF NEWGATE. Illustrated. New Edition. Demy 8vo, 16s.
MEMORIALS OF MILLBANK: or, Chapters in Prison History. With Illustrations. New Edition. Demy 8vo, 12s.

Grimble (A.).
SHOOTING AND SALMON FISHING: HINTS AND RECOLLECTIONS. With Illustrations. Second Edition. Demy 8vo, 16s.

Hall (Sidney).
A TRAVELLING ATLAS OF THE ENGLISH COUNTIES. Fifty Maps, coloured. Roan tuck, 10s. 6d.

Harris (Frank).
ELDER CONKLIN, AND OTHER STORIES. Crown 8vo.
[In the Press.

Hildebrand (Hans), *Royal Antiquary of Sweden.*
INDUSTRIAL ARTS OF SCANDINAVIA IN THE PAGAN TIMES. With numerous Woodcuts. Large crown 8vo, 2s. 6d.

Holmes (George C. V.).
MARINE ENGINES AND BOILERS. With 69 Woodcuts. Large crown 8vo, 3s.

Houssaye (Arsène).
BEHIND THE SCENES OF THE COMÉDIE FRANÇAISE, AND OTHER RECOLLECTIONS. Translated from the French. Demy 8vo, 14s.

Hovelacque (Abel).
THE SCIENCE OF LANGUAGE: LINGUISTIC, PHILOLOGY, AND ETYMOLOGY. With Maps. Large crown 8vo, 3s. 6d.

Hozier (H. M.).
TURENNE. With Portrait and Two Maps. Large crown 8vo, 4s.

Hudson (W. H.), C.M.Z.
IDLE DAYS IN PATAGONIA. With numerous Illustrations by J. SMIT and A. HARTLEY. Demy 8vo, 14s.
THE NATURALIST IN LA PLATA. With numerous Illustrations by J. SMIT. Second Edition. Demy 8vo, 16s.

Hueffer (F.).
HALF A CENTURY OF MUSIC IN ENGLAND. 1837-1887. Demy 8vo, 8s.

Hughes (W. R.), F.L.S.
A WEEK'S TRAMP IN DICKENS-LAND. With upwards of 100 Illustrations by F. G. KITTON, HERBERT RAILTON, and others. Demy 8vo, 16s.

Hutchinson (Rev. H. N.).
EXTINCT MONSTERS. A popular Account of some of the larger forms of Ancient Animal Life. With numerous Illustrations by J. SMIT and others, and a Preface by DR. HENRY WOODWARD, F.R.S. Demy 8vo. 12s.

Huntly (Marquis of).
TRAVELS, SPORTS, AND POLITICS IN THE EAST OF EUROPE. With Illustrations. Large crown 8vo, 12s.

INDUSTRIAL ARTS: Historical Sketches. With numerous Illustrations. Large crown 8vo, 3s.

Jackson (Frank G.).
DECORATIVE DESIGN. An Elementary Text Book of Principles and Practice. With numerous Illustrations. Second Edition. Large crown 8vo, 7s. 6d.

James (Henry A.), M.A.
HANDBOOK TO PERSPECTIVE. Crown 8vo, 2s. 6d.
PERSPECTIVE CHARTS, for use in Class Teaching. 2s.

Jokai (Maurus).
PRETTY MICHAL. Translated by R. NISBET BAIN. Crown 8vo, 5s.

Jones.
HANDBOOK OF THE JONES COLLECTION IN THE SOUTH KENSINGTON MUSEUM. With Portrait and Woodcuts. Large crown 8vo, 2s. 6d.

Jopling (Louise).
HINTS TO AMATEURS. A Handbook on Art. With Diagrams. Crown 8vo, 1s. 6d.

Junker (Dr. Wm.).
TRAVELS IN AFRICA. Translated from the German by Professor A. H. KEANE, F.R.G.S.
Volume I. DURING THE YEARS 1875 TO 1878. Containing 38 Full-page Plates and 125 Illustrations in the Text and Map. Demy 8vo, 21s.
Volume II. DURING THE YEARS 1879 TO 1882. Containing numerous Full-page Plates and Illustrations in the Text and Map. Demy 8vo, 21s.
Volume III. DURING THE YEARS 1882 TO 1886. Containing numerous Full-page Plates and Illustrations in the Text and Maps. Demy 8vo, 21s.

Kelly (James Fitzmaurice).
THE LIFE OF MIGUEL DE CERVANTES SAAVEDRA: A Biographical, Literary, and Historical Study, with a Tentative Bibliography from 1585 to 1892, and an Annotated Appendix on the "Canto de Caliope." Demy 8vo, 16s.

Kennard (Edward).
NORWEGIAN SKETCHES: FISHING IN STRANGE WATERS. Illustrated with 30 beautiful Sketches. Second Edition. Oblong folio, 21s. Smaller Edition, 14s.

LACORDAIRE'S JESUS CHRIST; GOD; AND GOD AND MAN. Conferences delivered at Notre Dame, in Paris. Seventh Thousand. Crown 8vo, 3s. 6d.

Laing (S.).
HUMAN ORIGINS: EVIDENCE FROM HISTORY AND SCIENCE. With Illustrations. Tenth Thousand. Demy 8vo, 3s. 6d.
PROBLEMS OF THE FUTURE AND ESSAYS. Tenth Thousand. Demy 8vo, 3s. 6d.
MODERN SCIENCE AND MODERN THOUGHT. . Fifteenth Thousand. Demy 8vo, 3s. 6d.
A MODERN ZOROASTRIAN. Seventh Thousand. Demy 8vo, 3s. 6d.

Lanin (E. B.).
RUSSIAN CHARACTERISTICS. Reprinted, with Revisions, from *The Fortnightly Review*. Demy 8vo, 14s.

Le Conte (Joseph).
EVOLUTION: ITS NATURE, ITS EVIDENCES, AND ITS RELATIONS TO RELIGIOUS THOUGHT. A New and Revised Edition. Crown 8vo, 6s.

Lefevre (André).
PHILOSOPHY, Historical and Critical. Translated, with an Introduction, by A. H. KEANE, B.A. Large crown 8vo, 3s. 6d.

Le Roux (H.).
ACROBATS AND MOUNTEBANKS. With over 200 Illustrations by J. GARNIER. Royal 8vo, 16s.

Leroy-Beaulieu (Anatole), *Member of the Institute of France.*
PAPACY, SOCIALISM, AND DEMOCRACY. Translated by B. L. O'DONNELL. Crown 8vo, 7s. 6d.

Leslie (R. C.).
THE SEA BOAT: HOW TO BUILD, RIG, AND SAIL HER. With Illustrations. Crown 8vo, 4s. 6d.
OLD SEA WINGS, WAYS, AND WORDS, IN THE DAYS OF OAK AND HEMP. With 135 Illustrations by the Author. Demy 8vo, 14s.
LIFE ABOARD A BRITISH PRIVATEER IN THE TIME OF QUEEN ANNE. Being the Journals of Captain Woodes Rogers, Master Mariner. Large crown 8vo, 9s.
A SEA-PAINTER'S LOG. With 12 Full-page Illustrations by the Author. Large crown 8vo, 12s.

Letourneau (Dr. Charles).
SOCIOLOGY. Based upon Ethnology. Demy 8vo, 3s. 6d.
BIOLOGY. With 83 Illustrations. A New Edition. Demy 8vo, 3s. 6d.

Lilly (W. S.).
ON SHIBBOLETHS. Demy 8vo, 12s.
ON RIGHT AND WRONG. Second Edition. Demy 8vo, 12s.
A CENTURY OF REVOLUTION. Second Edition. Demy 8vo, 12s.
CHAPTERS ON EUROPEAN HISTORY. 2 vols. Demy 8vo, 21s.
ANCIENT RELIGION AND MODERN THOUGHT. Second Edition. Demy 8vo, 12s.

Lineham (W. J.).
TEXT BOOK OF MECHANICAL ENGINEERING.
Part I. WORKSHOP PRACTICE.
Part II. THEORY AND EXAMPLES.
[In the Press.

Little (The Rev. Canon Knox).
THE CHILD OF STAFFERTON. Twelfth Thousand. Crown 8vo, boards, 1s.
THE BROKEN VOW. Seventeenth Thousand. Crown 8vo, boards, 1s.

Lloyd (W. W.), late 24th Regiment.
ON ACTIVE SERVICE. Printed in Colours. Oblong 4to, 5s.
SKETCHES OF INDIAN LIFE. Printed in Colours. 4to, 6s.

Malleson (Col. G. B.), C.S.I.
PRINCE EUGENE OF SAVOY. With Portrait and Maps. Large crown 8vo, 6s.
LOUDON. A Sketch of the Military Life of Gideon Ernest, Freicherr von Loudon. With Portrait and Maps. Large crown 8vo, 4s.

Mallock (W. H.).
A HUMAN DOCUMENT. New Edition, in One Volume. Crown 8vo, 3s. 6d.

Marceau (Sergent).
REMINISCENCES OF A REGICIDE. Edited from the Original MSS. of SERGENT MARCEAU, Member of the Convention, and Administrator of Police in the French Revolution of 1789. By M. C. M. SIMPSON. With Illustrations and Portraits. Demy 8vo, 14s.

Maskell (Alfred).
RUSSIAN ART AND ART OBJECTS IN RUSSIA. A Handbook to the Reproduction of Goldsmith's Work and Other Art Treasures. With Illustrations. Large crown 8vo, 4s. 6d.

Maskell (William).
IVORIES: ANCIENT AND MEDIÆVAL. With numerous Woodcuts. Large crown 8vo, 1s. 6d.
HANDBOOK TO THE DYCE AND FORSTER COLLECTIONS. With Illustrations. Large crown 8vo, 2s. 6d.

Maspéro (G.), late Director of Archæology in Egypt.
LIFE IN ANCIENT EGYPT AND ASSYRIA. Translated by A. P. Morton. With 188 Illustrations. Third Thousand. Crown 8vo, 5s.

Meredith (George).
(Works see page 16.)

Mills (John), *formerly Assistant to the Solar Physics Committee.*
ADVANCED PHYSIOGRAPHY (PHYSIOGRAPHIC ASTRO-NOMY). Designed to meet the Requirements of Students preparing for the Elementary and Advanced Stages of Physiography in the Science and Art Department Examinations, and as an Introduction to Physical Astronomy. Crown 8vo, 4s. 6d.
ELEMENTARY PHYSIOGRAPHIC ASTRONOMY. Crown 8vo, 2s. 6d
ALTERNATIVE ELEMENTARY PHYSICS. Crown 8vo, 1s. 6d.

Mills (John) and North (Barker).
QUANTITATIVE ANALYSIS (INTRODUCTORY LESSONS ON). With numerous Woodcuts. Crown 8vo, 1s. 6d.
HANDBOOK OF QUANTITATIVE ANALYSIS. Crown 8vo, 3s. 6d.

Mitre (General Don Bartolomé), *first President of the Argentine Republic.*
THE EMANCIPATION OF SOUTH AMERICA. Being a Condensed Translation, by WILLIAM PILLING, of "The History of San Martin." Demy 8vo, with Maps.

Molesworth (W. Nassau).
HISTORY OF ENGLAND FROM THE YEAR 1830 TO THE RESIGNATION OF THE GLADSTONE MINISTRY, 1874. Twelfth Thousand. 3 vols. Crown 8vo, 18s.
ABRIDGED EDITION. Large crown, 7s. 6d.

Nesbitt (Alexander).
GLASS. With numerous Woodcuts. Large crown 8vo, 2s. 6d.

O'Byrne (Robert), F.R.G.S.
THE VICTORIES OF THE BRITISH ARMY IN THE PENINSULA AND THE SOUTH OF FRANCE from 1808 to 1814. An Epitome of Napier's History of the Peninsular War, and Gurwood's Collection of the Duke of Wellington's Despatches. Crown 8vo, 5s.

Oliver (Professor D.), F.R.S., etc.
ILLUSTRATIONS OF THE PRINCIPAL NATURAL ORDERS OF THE VEGETABLE KINGDOM, prepared for the Science and Art Department, South Kensington. With 109 Plates. Oblong 8vo, plain, 16s.; coloured, £1 6s.

Oliver (E. E.), *Under-Secretary to the Public Works Department, Punjaub.*
ACROSS THE BORDER; or, PATHAN AND BILOCH. With numerous Illustrations by J. L. KIPLING, C.I.E. Demy 8vo, 14s.

Papus.
THE TAROT OF THE BOHEMIANS. The most ancient book in the world. For the exclusive use of the Initiates. An Absolute Key to Occult Science. With numerous Illustrations. Large crown 8vo, 7s. 6d.

Paske (Surgeon-General C. T.) and Aflalo (F. G.).
THE SEA AND THE ROD. With Illustrations. Crown 8vo, 4s. 6d.

Paterson (Arthur).
A PARTNER FROM WEST. A Novel. Crown 8vo, 5s.

Payton (E. W.).
ROUND ABOUT NEW ZEALAND. Being Notes from a Journal of Three Years' Wandering in the Antipodes. With Twenty Original Illustrations by the Author. Large crown 8vo, 12s.

Pelagius.
HOW TO BUY A HORSE. With Hints on Shoeing and Stable Management. Crown 8vo, 1s.

Pierce (Gilbert).
THE DICKENS DICTIONARY. A Key to the Characters and Principal Incidents in the Tales of Charles Dickens. New Edition. Large crown, 5s.

Perrot (Georges) and Chipiez (Chas.).

A HISTORY OF ANCIENT ART IN
PERSIA. With 254 Illustrations
and 12 Steel and Coloured Plates.
Imperial 8vo, 21s.

A HISTORY OF ANCIENT ART IN
PHRYGIA—LYDIA, CARIA, AND
LYCIA. With 280 Illustrations.
Imperial 8vo, 15s.

A HISTORY OF ANCIENT ART IN
SARDINIA, JUDÆA, SYRIA, AND
ASIA MINOR. With 395 Illustra-
tions. 2 vols. Imperial 8vo, 36s.

A HISTORY OF ANCIENT ART IN
PHŒNICIA AND ITS DEPEN-
DENCIES. With 654 Illustrations.
2 vols. Imperial 8vo, 42s.

A HISTORY OF ART IN CHALDÆA
AND ASSYRIA. With 452 Illustrations. 2 vols. Imperial 8vo, 42s.

A HISTORY OF ART IN ANCIENT
EGYPT. With 600 Illustrations. 2
vols. Imperial 8vo, 42s.

Pollen (J. H.).

GOLD AND SILVER SMITH'S
WORK. With numerous Woodcuts.
Large crown 8vo, 2s. 6d.

ANCIENT AND MODERN FURNI-
TURE AND WOODWORK. With
numerous Woodcuts. Large crown
8vo, 2s. 6d.

Poole (Stanley Lane), B.A., M.R.A.S.

THE ART OF THE SARACENS IN
EGYPT. Published for the Committee of Council on Education.
With 108 Woodcuts. Large crown
8vo, 4s.

Poynter (E. J.), R.A.

TEN LECTURES ON ART. Third
Edition. Large crown 8vo, 9s.

Pratt (Robert).

SCIOGRAPHY, OR PARALLEL
AND RADIAL PROJECTION OF
SHADOWS. Being a Course of
Exercises for the use of Students in
Architectural and Engineering Drawing, and for Candidates preparing for
the Examinations in this subject and
in Third Grade Perspective. Oblong
quarto, 7s. 6d.

QUEEN OF SPADES, THE, and
OTHER STORIES. With a Bio-
graphy. Translated from the Russian
by MRS. SUTHERLAND EDWARDS.
Illustrated. Crown 8vo, 3s. 6d.

Rae (W. Fraser).

AUSTRIAN HEALTH RESORTS
THROUGHOUT THE YEAR.
A New and Enlarged Edition. Crown
8vo, 5s.

RAPHAEL ; his Life, Works, and Times.
By EUGENE MUNTZ. Illustrated
with about 200 Engravings. A New
Edition, revised from the Second
French Edition. By W. ARMSTRONG,
B.A. Imperial 8vo, 25s.

Redgrave (Gilbert).

OUTLINES OF HISTORIC OR-
NAMENT. Translated from the
German. Edited by GILBERT RED-
GRAVE. With numerous Illustra-
tions. Crown 8vo, 4s.

Redgrave (Richard), R.A.

MANUAL OF DESIGN. With
Woodcuts. Large crown 8vo, 2s. 6d.

ELEMENTARY MANUAL OF
COLOUR, with a Catechism on
Colour. 24mo, cloth, 9d.

Redgrave (Samuel).

A DESCRIPTIVE CATALOGUE
OF THE HISTORICAL COL-
LECTION OF WATER-COLOUR
PAINTINGS IN THE SOUTH
KENSINGTON MUSEUM. With
numerous Chromo-lithographs and
other Illustrations. Royal 8vo, £1 1s.

Renan (Ernest).
THE FUTURE OF SCIENCE: Ideas of 1848. Demy 8vo, 18s.
HISTORY OF THE PEOPLE OF ISRAEL.
FIRST DIVISION. Till the time of King David. Demy 8vo, 14s.
SECOND DIVISION. From the Reign of David up to the Capture of Samaria. Demy 8vo, 14s.
THIRD DIVISION. From the time of Hezekiah till the Return from Babylon. Demy 8vo, 14s.
RECOLLECTIONS OF MY YOUTH. Translated from the French, and revised by MADAME RENAN. Second Edition. Crown 8vo, 3s. 6d.

Riaño (Juan F.).
THE INDUSTRIAL ARTS IN SPAIN. With numerous Woodcuts. Large crown 8vo, 4s.

Roberts (Morley).
IN LOW RELIEF: A Bohemian Transcript. Second Edition. Crown 8vo, 3s. 6d.

Robson (George).
ELEMENTARY BUILDING CONSTRUCTION. Illustrated by a Design for an Entrance Lodge and Gate. 15 Plates. Oblong folio, sewed, 8s.

Rock (The Very Rev. Canon), D.D.
TEXTILE FABRICS. With numerous Woodcuts. Large crown 8vo, cloth, 2s. 6d.

Roosevelt (Blanche).
ELIZABETH OF ROUMANIA. A Study. With Two Tales from the German of Carmen Sylva, Her Majesty Queen of Roumania. With Two Portraits and Illustration. Demy 8vo, 12s.

Ross (Mrs. Janet).
EARLY DAYS RECALLED. With Illustrations and Portrait. Crown 8vo, 5s.

Russan (Ashmore) and Boyle (Fredk.).
THE ORCHID SEEKERS: a Story of Adventure in Borneo. Illustrated by ALFRED HARTLEY. Crown 8vo.

Schauermann (F. L.).
WOOD-CARVING IN PRACTICE AND THEORY, AS APPLIED TO HOME ARTS. Containing 124 Illustrations. Second Edition. Large crown 8vo, 5s.

Schreiner (Olive) (Ralph Iron).
THE STORY OF AN AFRICAN FARM. Crown 8vo, 1s.; in cloth, 1s. 6d.
A New Edition, on Superior Paper, and strongly bound in cloth. Crown 8vo, 3s. 6d.

Scott (Leader).
THE RENAISSANCE OF ART IN ITALY: an Illustrated Sketch. With upwards of 200 Illustrations. Medium 4to, 18s.

Seeman (O.).
THE MYTHOLOGY OF GREECE AND ROME, with Special Reference to its Use in Art. From the German. Edited by G. H. BIANCHI. 64 Illustrations. New Edition. Crown 8vo, 5s.

Seton Karr (H. W.), F.R.G.S., etc.
BEAR HUNTING IN THE WHITE MOUNTAINS; or, Alaska and British Columbia Revisited. Illustrated. Large crown, 4s. 6d.
TEN YEARS' TRAVEL AND SPORT IN FOREIGN LANDS; or, Travels in the Eighties. Second Edition, with additions and Portrait of Author. Large crown 8vo, 5s.

Shirreff (Emily).
A SHORT SKETCH OF THE LIFE OF FRIEDRICH FRÖBEL; a New Edition, including Fröbel's Letters from Dresden and Leipzig to his Wife, now first Translated into English. Crown 8vo, 2s.
HOME EDUCATION IN RELATION TO THE KINDERGARTEN. Two Lectures. Crown 8vo, 1s. 6d.

Simkin (R.).
LIFE IN THE ARMY: Every-day Incidents in Camp, Field, and Quarters. Printed in Colours. Oblong 4to, 5s.

Simmonds (T. L.).
ANIMAL PRODUCTS: their Preparation, Commercial Uses and Value. With numerous Illustrations. Large crown 8vo, 3s. 6d.

Sinnett (A. P.).
ESOTERIC BUDDHISM. Annotated and enlarged by the Author. Seventh Edition. Crown 8vo, 3s. 6d.
KARMA. A Novel. New Edition. Crown 8vo, 3s.

Smith (Major R. Murdock), R.E.
PERSIAN ART. With Map and Woodcuts. Second Edition. Large crown 8vo, 2s.

Statham (H. H.).
MY THOUGHTS ON MUSIC AND MUSICIANS. Illustrated with Frontispiece and Musical Examples. Demy 8vo, 18s.

Stoddard (C. A.).
SPANISH CITIES: with Glimpses of Gibraltar and Tangiers. With 18 Illustrations. Large crown 8vo, 7s. 6d.
ACROSS RUSSIA FROM THE BALTIC TO THE DANUBE. With numerous Illustrations. Large crown 8vo, 7s. 6d.

Stokes (Margaret).
EARLY CHRISTIAN ART IN IRELAND. With 106 Woodcuts. Demy 8vo, 7s. 6d.
Cheaper Edition, crown 8vo, 4s.

STORIES FROM "BLACK AND WHITE." By THOMAS HARDY, J. M. BARRIE, W. CLARK RUSSELL, W. E. NORRIS, JAMES PAYN, GRANT ALLEN, MRS. LYNN LINTON, and MRS. OLIPHANT. With numerous Illustrations. Crown 8vo, 5s.

Sutcliffe (John).
THE SCULPTOR AND ART STUDENT'S GUIDE to the Proportions of the Human Form, with Measurements in feet and inches of full-grown Figures of both Sexes, and of various Ages. By DR. G. ZCHADOW. Plates reproduced by J. SUTCLIFFE. Oblong folio, 31s. 6d.

SUVOROFF, LIFE OF. By LIEUT.-COL. SPALDING. Crown 8vo, 6s.

Symonds (John Addington).
ESSAYS, SPECULATIVE AND SUGGESTIVE. New Edition.
[In the Press.

Taine (H. A.).
NOTES ON ENGLAND. With Introduction by W. FRASER RAE. Eighth Edition. With Portrait. Crown 8vo, 5s.

Tanner (Professor), F.C.S.
HOLT CASTLE; or, Threefold Interest in Land. Crown 8vo, 4s. 6d.
JACK'S EDUCATION; OR, HOW HE LEARNT FARMING. Second Edition. Crown 8vo, 3s. 6d.

Taylor (Edward R.).
ELEMENTARY ART TEACHING: An Educational and Technical Guide for Teachers and Learners, including Infant School-work; The Work of the Standards; Freehand; Geometry; Model Drawing; Nature Drawing; Colours; Light and Shade; Modelling and Design. With over 600 Diagrams and Illustrations. Large crown 8vo, 10s. 6d.

Temple (Sir R.), Bart., M.P., G.C.S.I.
COSMOPOLITAN ESSAYS. With Maps. Demy 8vo, 16s.

Thomson (D. C.).
THE BARBIZON SCHOOL OF PAINTERS: Corot, Rousseau, Diaz, Millet, and Daubigny. With 130 Illustrations, including 36 Full-page Plates, of which 18 are Etchings. 4to, cloth, 42s.

Topinard (Dr. Paul).
ANTHROPOLOGY. With a Preface by PROFESSOR PAUL BROCA. With 49 Illustrations. Demy 8vo, 3s. 6d.

Traherne (Major).
THE HABITS OF THE SALMON. Crown 8vo, 3s. 6d.

Traill (H. D.).
THE NEW LUCIAN. Being a Series of Dialogues of the Dead. Demy 8vo, 12s.

Trollope (Anthony).
THE CHRONICLES OF BARSET-SHIRE. A Uniform Edition in 8 vols., large crown 8vo, handsomely printed, each vol. containing Frontispiece. 6s. each.
THE WARDEN AND BARCHESTER TOWERS. 2 vols.
DR. THORNE.
FRAMLEY PARSONAGE.
THE SMALL HOUSE AT ALLINGTON. 2 vols.
LAST CHRONICLE OF BARSET. 2 vols.

Troup (J. Rose).
WITH STANLEY'S REAR COLUMN. With Portraits and Illustrations. Second Edition. Demy 8vo, 16s.

Underhill (G. F.).
IN AND OUT OF THE PIG SKIN. With Illustrations. Crown 8vo, 1s.

Veron (Eugene).
ÆSTHETICS. Translated by W. H. ARMSTRONG. Large crown 8vo, 3s. 6d.

Walford (Major), R.A.
PARLIAMENTARY GENERALS OF THE GREAT CIVIL WAR. With Maps. Large crown 8vo, 4s.

Walker (Mrs.).
UNTRODDEN PATHS IN ROUMANIA. With 77 Illustrations. Demy 8vo, 10s, 6d.
EASTERN LIFE AND SCENERY, with Excursions to Asia Minor, Mitylene, Crete, and Roumania. 2 vols., with Frontispiece to each vol. Crown 8vo, 21s.

Wall (A.).
A PRINCESS OF CHALCO. A Novel. With Illustrations. Crown 8vo, 6s.

Ward (James).
ELEMENTARY PRINCIPLES OF ORNAMENT. With 122 Illustrations in the text. 8vo, 5s.
THE PRINCIPLES OF ORNAMENT. Edited by GEORGE AITCHISON, A. R. A., Professor of Architecture at the Royal Academy of Arts. 8vo, 7s. 6d.

Watson (John).
POACHERS AND POACHING. With Frontispiece. Crown 8vo, 7s. 6d.
SKETCHES OF BRITISH SPORTING FISHES. With Frontispiece. Crown 8vo, 3s. 6d.

White (Walter).
A MONTH IN YORKSHIRE. With a Map. Fifth Edition. Post 8vo, 4s.
A LONDONER'S WALK TO THE LAND'S END. With 4 Maps. Third Edition. Post 8vo, 4s.

Woodgate (W. B.).
A MODERN LAYMAN'S FAITH. Demy 8vo.   [*In February.*

Wornum (R. N.).
ANALYSIS OF ORNAMENT: THE CHARACTERISTICS OF STYLES. With many Illustrations. Ninth Edition. Royal 8vo, cloth, 8s.

Worsaae (J. J. A.).
INDUSTRIAL ARTS OF DENMARK, FROM THE EARLIEST TIMES TO THE DANISH CONQUEST OF ENGLAND. With Maps and Woodcuts. Large crown 8vo, 3s. 6d.

Wotton (Mabel E.).
A GIRL DIPLOMATIST. Crown 8vo, 3s. 6d.

Wrightson (Prof. John), *President of the College of Agriculture, Downton.*
PRINCIPLES OF AGRICULTURAL PRACTICE OF AN INSTRUCTIONAL SUBJECT. With Geological Map. Second Edition. Crown 8vo, 5s.
FALLOW AND FODDER CROPS. Crown 8vo, 5s.

CHARLES DICKENS'S WORKS.

CROWN EDITION COMPLETE IN 17 VOLS.,

Printed on good paper, from type specially cast for this Edition, and containing

All the Illustrations by Seymour, Phiz (H. K. Browne), Tenniel, Leech, Landseer, Cattermole, Cruikshank, Marcus Stone, Luke Fildes, and others.

PRICE FIVE SHILLINGS EACH VOLUME.

The Pickwick Papers. With Forty-three Illustrations by Seymour and Phiz.

Nicholas Nickleby. With Forty Illustrations by Phiz.

Dombey and Son. With Forty Illustrations by Phiz.

David Copperfield. With Forty Illustrations by Phiz.

Sketches by Boz. With Forty Illustrations by George Cruikshank.

Martin Chuzzlewit. With Forty Illustrations by Phiz.

The Old Curiosity Shop. With Seventy-five Illustrations by George Cattermole and H. K. Browne.

Barnaby Rudge: a Tale of the Riots of 'Eighty. With Seventy-eight Illustrations by George Cattermole and H. K. Browne.

Oliver Twist and **A Tale of Two Cities.** With Twenty-four Illustrations by Cruikshank, and Sixteen by Phiz.

Bleak House. With Forty Illustrations by Phiz.

Little Dorrit. With Forty Illustrations by Phiz.

Our Mutual Friend. With Forty Illustrations by Marcus Stone.

American Notes; Pictures from Italy; and **A Child's History of England.** With Sixteen Illustrations by Marcus Stone.

Christmas Books and **Hard Times.** With Sixty-seven Illustrations by Landseer, Maclise, Stanfield, Leech, Doyle, F. Walker, etc.

Christmas Stories and Other Stories, including **Humphrey's Clock.** With Illustrations by Dalziel, Charles Green, Mahoney, Phiz, Cattermole, etc.

Great Expectations; Uncommercial Traveller. With Sixteen Illustrations by Marcus Stone.

Edwin Drood and **Reprinted Pieces.** With Sixteen Illustrations by Luke Fildes and F. Walker.

Uniform with above in size and binding.

The Life of Charles Dickens. By JOHN FORSTER. With Portraits and Illustrations. Added at the request of numerous subscribers.

The Dickens Dictionary: a Key to the Characters and Principal Incidents in the Tales of Charles Dickens.

The Lazy Tour of Two Idle Apprentices; No Thoroughfare; The Perils of Certain English Prisoners. By CHARLES DICKENS and WILKIE COLLINS. With Illustrations. Crown 8vo, 5s.

** These Stories are now reprinted in complete form for the first time.

CHARLES DICKENS'S WORKS.

HALF-CROWN EDITION.

This Edition will contain the whole of Dickens's Works, with all the Illustrations, and be complete in about 19 *or* 20 *Crown 8vo Volumes.*

Printed from the Edition that was carefully corrected by the Author in 1867 and 1868.

PRICE TWO SHILLINGS AND SIXPENCE EACH VOLUME.

The Pickwick Papers. With Forty-three Illustrations by Seymour and Phiz.

Barnaby Rudge : a Tale of the Riots of 'Eighty. With Seventy-six Illustrations by George Cattermole and H. K. Browne.

Oliver Twist. With Twenty-four Illustrations by George Cruikshank.

The Old Curiosity Shop. With Seventy-five Illustrations by George Cattermole and H. K. Browne.

David Copperfield. With Forty Illustrations by Phiz.

Nicholas Nickleby. With Forty Illustrations by Phiz.

Martin Chuzzlewit. With Forty Illustrations by Phiz.

Dombey and Son. With Forty Illustrations by Phiz.

Christmas Books. With Sixty-three Illustrations by Landseer, Doyle, Maclise, Leech, etc.

Sketches by Boz. With Forty Illustrations by George Cruikshank.

Bleak House. With Forty Illustrations by Phiz.

Little Dorrit. With Forty Illustrations by Phiz.

THE ILLUSTRATED LIBRARY EDITION.

Complete in 30 vols., with the Original Illustrations, demy 8vo, 10s. each; or Sets, £15.

LIBRARY EDITION.

Complete in 30 vols., with the Original Illustrations, post 8vo, 8s. each; or Sets, £12.

THE POPULAR LIBRARY EDITION.

In 30 vols., large crown 8vo, £6; separate volumes, 4s. each.

THE "CHARLES DICKENS" EDITION.

In crown 8vo, in 21 vols., cloth, with Illustrations, £3 16s.

THE CABINET EDITION.

In 32 vols., small fcap. 8vo, marble paper sides, cloth backs, with uncut edges, 1s. 6d. each. Each Volume contains 8 Illustrations reproduced from the Originals.

THOMAS CARLYLE'S WORKS.

THE ASHBURTON EDITION.

An entirely New Edition, handsomely printed, containing all the Portraits and Illustrations; in 17 vols., demy 8vo, 8s. each.

CHEAP AND UNIFORM EDITION.

23 vols., crown 8vo, cloth, £7 5s.

LIBRARY EDITION.

Handsomely printed in 34 vols., demy 8vo, cloth, £15 3s.

PEOPLE'S EDITION.

37 vols., small crown 8vo, 37s.; separate vols., 1s. each.

Sartor Resartus. With Portrait of Thomas Carlyle.
French Revolution: a History. 3 vols.
Oliver Cromwell's Letters and Speeches. 5 vols. With Portrait of Oliver Cromwell.
On Heroes and Hero Worship and the Heroic in History.
Past and Present.
Critical and Miscellaneous Essays. 7 vols.
The Life of Schiller, and Examination of His Works. With Portrait.
Latter-Day Pamphlets.
Wilhelm Meister. 3 vols.
Life of John Sterling. With Portrait.
History of Frederick the Great. 10 vols.
Translations from Musæus, Tieck, and Richter. 2 vols.
The Early Kings of Norway; Essay on the Portrait of Knox.

Or in Sets, 37 vols. in 18, 37s.

GEORGE MEREDITH'S WORKS.

A New and Uniform Edition. Crown 8vo, 3s. 6d. each.
Copies of the Six-Shilling Edition are still to be had.

One of Our Conquerors.
Diana of the Crossways.
Evan Harrington.
The Ordeal of Richard Feverel.
The Adventures of Harry Richmond.
Sandra Belloni.
Vittoria.
Rhoda Fleming.
Beauchamp's Career.
The Egoist.
The Shaving of Shagpat; and Farina.

CHARLES DICKENS AND EVANS, CRYSTAL PALACE PRESS.

www.ingramcontent.com/pod-product-compliance
Lightning Source LLC
Chambersburg PA
CBHW021402210326
41599CB00011B/980